Upholstery
S T Y L E S

A DESIGN SOURCEBOOK

GILLIAN WALKLING

VNR VAN NOSTRAND REINHOLD
_____ New York

A QUARTO BOOK

Copyright © 1989 by Quarto Publishing plc

ISBN 0-442-23844-4

This book was designed and produced by
Quarto Publishing plc
The Old Brewery
6 Blundell Street
London N7 9BH

Published in the U.S.A. by
Van Nostrand Reinhold
115 Fifth Avenue
New York, New York 10003

Distributed in Canada by
Macmillan of Canada
Division of Canada Publishing Corporation
164 Commander Boulevard
Agincourt, Ontario M1S 3C7. Canada

PROJECT EDITOR: Charyn Jones
DESIGNER: Graham Davis

SENIOR EDITOR: Kate Kirby

ARTISTS: David Kemp, Rob Stone

CALLIGRAPHER: Heleen Franken

PICTURE RESEARCHER: Joanna Wiese

ART DIRECTOR: Moira Clinch
EDITORIAL DIRECTOR: Carolyn King

Typeset by Burbeck Associates Ltd, Harlow, Essex.
Manufactured in Hong Kong by
Regent Publishing Services Ltd.
Printed by Lee Fung Asco Printers Ltd, Hong Kong

#ANF 10-12-12

16   15   14   13   12   11   10   9   8   7   6   5   4   3   2   1

*Right: A brand new
Victorian-style chaise
longue.*

*Previous page: A 1925
Sue et Mare salon suite
covers by Dufrêne*

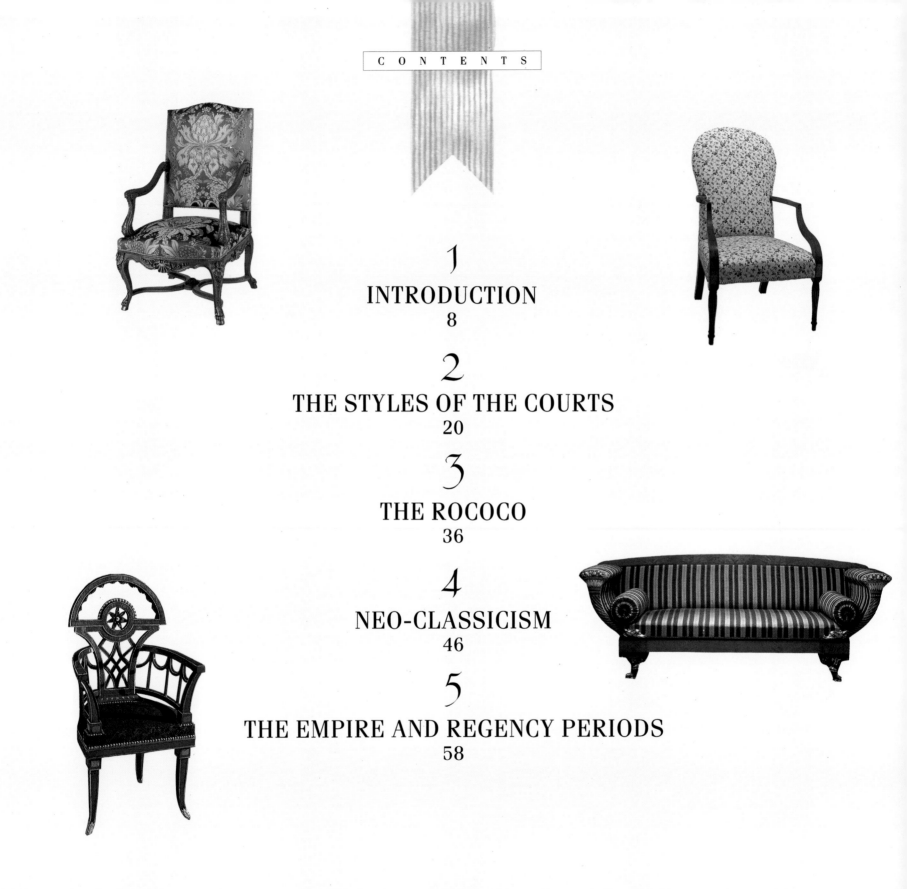

INTRODUCTION

LOOKING AT THE PAST

·

FURNITURE IN HISTORY

·

THE UPHOLSTERER'S ART

·

TEXTILES AND TRIMMINGS

·

RESTORATION

*A model of a sleeping
chamber, Ham House,
Surrey, England.*

LOOKING AT THE PAST

Although the interiors of houses throughout history have reflected domestic habits and social mores, many of the furnishings could be seen as peripheral; they were not absolutely essential and did not always serve a specific function. Today we live in houses with a huge diversity of interior decoration and soft furnishings. Indeed the trend to minimalism sometimes proscribes furnishings — even paintings are kept in cupboards and only brought out for viewing now and again.

The other interest in furniture lies in the level of craftsmanship and the technology involved in the manufacture — the easy chair, for example, derived from an invalid chair which was equipped with hinges to allow the invalid to lie in a semi-reclining position. Many modern chairs, on the other hand, are intended to be admired for their form as well as their function; they are pieces of sculpture.

The lives of our predecessors have always been a source of fascination, and the furnishings they chose to use for specific purposes or to decorate their dwellings is part of that fascination. Since upholstered furniture first appeared in the late sixteenth century, it has gradually diminished in importance. The social values which the chair, for example, initially reflected have radically changed; although it may still be used as a focus in the room, its function is to provide somewhere to sit.

· AUTHENTIC PERIOD DESIGN ·

The viewing of historic houses and their contents has become a popular leisure pursuit. Numerous organizations throughout the United States and Europe are working to preserve historic buildings and are painstakingly restoring or re-creating authentic period interiors. A wealth of new periodicals has appeared which are designed to show how authentic period effects can be created, even in a modern house, by careful choice of furnishings.

One of the most striking aspects of historic houses is the abundant use of textiles as a source of decoration, not just for curtains and upholstered furniture, but for wall hangings and elaborately draped beds too. The right choice of fabric and trimmings and the correct shape of the stuffing are of fundamental

10

The nineteenth century saw a proliferation of journals, such as House Beautiful (left), aimed at the middle-classes.

Pattern books often provided details of fabric designs as well as the furniture itself (above).

This engraving (below), from Diderot's Encyclopédie (published 1751-1772), gives an idea of contemporary patterns.

The survival of furnishings in their original settings is an invaluable source of information. This room (right) is at Osterley Park in London.

importance to the design of a piece of furniture as a whole, yet surprisingly, until recently, the subject of upholstery has been largely ignored.

The most obvious reason for this neglect is lack of surviving practical evidence; very few pieces of upholstered furniture pre-dating about 1850 have survived intact. Textiles are particularly vulnerable to the effects of dirt, light and insects, and simple wear and tear and changes in fashion have led to their disappearance. In view of the scarcity of surviving examples, information about early upholstery has to be sought elsewhere. Details of the type, cost and source of materials can often be found in old inventories, bills, letters and literature, but the most telling documents are paintings and engravings, where the depth and shape of the stuffing, as well as the trimmings and

sometimes the fabric's pattern, can clearly be seen. They can also give a clue as to how the furniture was used.

Design manuals and cabinet-makers' pattern books, which first appeared in the late seventeenth century, can also give an indication of the correct line of early upholstery, although such details were not always included and certainly very few publications made suggestions for appropriate types of fabric and trimmings.

One of the pitfalls of this type of research is that it tends to give a distorted view of the subject as a whole. Most of the available information comes from royal palaces and the grand houses of the wealthy, where continuous family or state ownership has helped to preserve the contents. While these documents clearly show the development of fashionable design, it has to

be remembered that the opulent and up-to-date furnishings these buildings contained were the property of the privileged few and were largely for show. The average householder would have been content with something made along the same lines, but much simpler in terms of materials.

In medieval Europe, rich textiles were a symbol of wealth and status and were prominently displayed when their owner wanted to impress those around him. The chair too was a symbol of power and was only used by the master of a household, everyone else, including his wife, having to be content with stools or benches. Comfort was provided by loose cushions. By the sixteenth century this practice had disappeared and chairs were used in households of all sizes.

FURNITURE IN HISTORY

uring the first half of the seventeenth century the layout of houses was governed by rigid social conventions. Grand houses had two suites of rooms — the state rooms, which were designed for visiting royalty and nobles, and the family rooms for normal domestic use. The state rooms were principally for show and were used for formal entertaining. The rooms of both suites were arranged in sequence, beginning with a reception room or ante-chamber, and gradually increasing in importance until the final and most significant room, the bedchamber, was reached. The bed itself was a symbol of power and both social and business callers were received in the bedroom.

The arrangement of furniture in all rooms was strictly formal. Long sets of matching chairs stood in rows against the wall and were only positioned in the middle of the room when in use, a practice which is assumed to account for the total absence of decoration on the backs of chairs. As they were only seen by servants, whose opinions didn't matter, there was no need for any ornament.

Upholstered backless stools or *tabourets* were commonly used by women. A refinement of these *tabourets* was the back stool or farthingale chair, and by 1600 these frequently had fixed upholstery of some sort on the back and seat which was trimmed with widely spaced nails and colored fringes. The short length of upright visible between the back and seat was covered with material and the gap made it comfortable for women wearing the farthingale, the

THE X-FRAME CHAIR

The grandest chair, or chair of State, in the seventeenth century was usually of X-frame construction; the back and seat had strips of material slung across the frame without any form of stuffing and the whole chair was completely covered in rich fabric bordered with closely spaced gilt-headed nails and decorated with gold and silver fringes. The folding X-frame had been popular in medieval times because of its collapsible design.

*A*n X-frame chair of State (dated 1610).

*A*s a symbol of power, the State bed (left) was the most elaborate and valuable piece of furniture in a grand house.

*T*he most common type of upholstered chair in a seventeenth-century house was the back-stool (left), literally, a stool with a back.

padded hip undergarment that held out skirts. In France there were two versions of these stools, one high and one low. The low types were for women and were referred to as *chaises à demoiselle* or *caquetoires* (*caqueter* meaning to chatter).

The chair coverings were usually *en suite* with the wall and window hangings and with the bed also. Sometimes a table was draped with a carpet of the same fabric.

THE EIGHTEENTH AND NINETEENTH CENTURIES ·

The formal arrangement of furniture persisted in Europe throughout the eighteenth century, as did, to a large extent, the use of co-ordinated fabrics. It was not until about 1810 that furniture began to take up a permanent position in the middle of the room. This move reflected the slightly less formal lifestyle and an increase in the amount of time a family spent together during the day. Presumably it was simply easier to leave the furniture standing where it was wanted, rather than have the servants constantly · moving it backwards and forwards. Long sets of identical chairs began to disappear and greater

numbers and varieties of occasional chairs were introduced into the room.

Although the same conventions governed both France and England in the eighteenth century, it is interesting to see how national characteristics affected the type of furniture produced. Despite the dominance of France over England where design was concerned, in England comfort was obviously of secondary importance to appearance. The way of living made few allowances for relaxation. In pleasure-loving France, however, the number

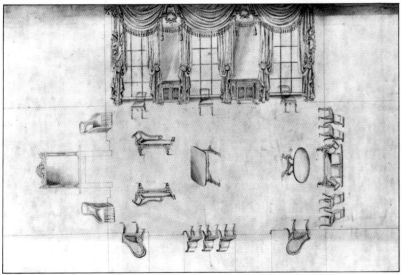

This set of tapestry seat covers and wall hangings at Osterley Park (above), with designs after Boucher, is one of several specially produced in the 1770s at the Gobelins Tapestry Works in Paris.

In England the upholsterer acted as an interior designer as well as a supplier of furnishings. In this 1820s Gillows' scheme for a drawing room (left) the arrangement of furniture is beginning to take on a less formal look.

and variety of upholstered chairs, sofas and daybeds were far greater and comfort was an important consideration.

· INFLUENTIAL FACTORS ·

It is of course impossible to make too many generalizations about upholstery in Europe as a whole. In some countries it was given far more prominence than in others and similarly, within an individual country, at some times it was more important than at others. In France, for example, where fashionable furniture was

highly sophisticated, upholstery played an important role until about 1820. French influence was strongly felt throughout Europe and particularly in Britain. In turn, English designers of the eighteenth century exerted a strong influence on early American furniture makers, many of whom were of British origin.

Because of understandable practical difficulties in obtaining both materials and up-to-date design information, until about 1750 American furniture lagged some years behind that in Europe, but soon caught up and developed its own characteristic forms. Although largely based on English design in the eighteenth century, early nineteenth-century American furniture shows strong French influence following the emigration of a large number of craftsmen from France after the Revolution. A century later, French design shown at the International Exhibition in Paris in 1925 exerted a strong influence in America and Europe.

The most difficult period to define with any accuracy is that covered by the years of Queen Victoria's reign, about 1840-1900. Ironically, although the Victorians are famous for their rather over-stuffed, deep-buttoned upholstery, in terms of design, upholstery was far less significant that at any previous time. Mechanization and mass-production combined to produce ranges of average-quality upholstery that was well suited to the demands of the increasingly prosperous, but not particularly discerning, middle-class population.

THE UPHOLSTERER'S ART

Traditional methods of upholstery are still used today despite the invention of synthetic fibers and a number of mechanical labor-saving devices. Upholstery is a skilled craft requiring judgement and dexterity and every piece of furniture needs to be approached differently. The stuffing is built up gradually following the contours of the frame and the upholsterer has to judge whether the shape and height look and feel right.

Early upholstery was rudimentary. Girthweb — the type of webbing used for making saddle girths — was woven across the frame and tacked at the edges. In France it was close-woven, but elsewhere there were large gaps between the girths. The webbing was then covered with a piece of hessian or coarse sackcloth and some form of padding was piled on, either feathers, wool, animal hair, or vegetable matter such as straw, chaff, rushes or dried grasses and leaves. Occasionally the padding was roughly secured with bridle ties before the top cover was nailed on. The backs of chairs were usually stuffed with washed and curled horsehair, which was more expensive than other padding but stayed in place better, stitched onto a linen or canvas backing. After about 1660 horsehair was used to a much greater extent and the front and sometimes the sides of the seat, which received the greatest wear, were strengthened with an extra band of stuffing sewn into a roll of hessian or canvas. An additional layer of fabric, usually linen, was placed between the horsehair and the top cover to prevent the hair from working its way through the fine fabric above. Some chairs had loose cushions filled with down or feathers.

In the eighteenth century these techniques were refined and complicated curved and straight-edged upholstery was carefully molded by more precise stuffing and stitching. Often there were two layers of stuffing on the seat — one thin layer of horsehair above a thicker layer of cheap vegetable fiber. Sometimes the padding was held in place with stitching or with regularly spaced ties which gave the surface of the upholstery a slightly indented pattern. This is now called shallow buttoning or tufting because silk tufts and not buttons were used as a decorative finish to each tie. In France, tufting was described as

The traditional appearance of this comfortable easy chair (above) belies its modern upholstery methods and materials and the clever reclining mechanism.

The owner of this eighteenth-century chair (right) by the sculptor Brustolon may have spent more on the upholstery than the frame.

capitonné. Buttons did not replace the ties until the end of the eighteenth century.

Until the middle of the eighteenth century, the upholstery of a chair or sofa was its most important part and often cost more in materials and workmanship than the frame. Consequently the upholsterer, or "upholder," was held in high esteem. In France the two crafts of *tapissier* (upholsterer) and *menuisier* (chair maker) were kept entirely separate. In England the two were sometimes combined.

The upholsterer was not just a supplier of upholstered furniture. He advised on and supplied all the textile furnishings of a house, from the elaborate wall hangings and curtains down to simple pillows and mattresses. As an interior designer, advising his clients on their choice

14

RE-WEBBING A SEAT FRAME

The webbing is always put on in the same way whether it goes under the frame on a sprung seat or over the top of it. Never cut the webbing into lengths before you start; work from the roll using the tack holes of the old web as a guide.

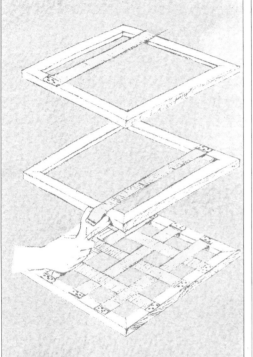

Fold the end of the webbing over 3cm (1in) and secure it with four or five 15mm (⅝in) tacks about 12mm (½in) in from the edge of the rail. Using a special webbing stretcher, pull the webbing until it is taut and firm but not over-strained and hammer in three tacks before releasing the stretcher. Cut the webbing 3cm (1in) from the tacks, fold over the end and secure with two more tacks.

Complete the procedure in one direction before beginning the next and always weave the cross webbing in before you tack the far end. In a sprung seat, straighten up the springs and sew them to the webbing with twine at intervals.

and arrangement of furnishings, he probably had a greater influence on the final appearance of a house than the architect who designed it. (Most upholsterers were also undertakers and supplied the accessories that decorated the funeral cortège and the elaborate black drapes and ornaments that decorated a house during mourning.)

THE COILED SPRING ·

Deep buttoning resulted from the widespread use of the coiled spring after 1820, which greatly deepened the padding. This springing was the only major advance in upholstery technique for over a century. The addition of a layer of springs underneath the layers of stuffing allowed the upholstery to move slightly and therefore made it more comfortable. The backs as well as the seats were often sprung and the thicker stuffing needed to cover the springs was prevented from slipping by the deep-set buttons. The resulting pleating of the material became a decorative feature and buttoning

sometimes appeared on the arms too, even though they were not always themselves sprung.

At first springs were attached at the bottom to wooden planks, similar to modern ready-made wood and metal sprung units, which simply slot into the frame. These were soon replaced by more flexible webbing.

Between 1650 and 1800, horsehair was used for the padding, but as it became more expensive, other materials were used to pad it out. These have included coconut fiber, known as coir or kerly fiber, various coarse grasses and leaves, cotton, rags and wool waste, and even wood chippings. This century has seen the introduction of rubber webbing, staple guns and latex and polyurethane foam, although they are really only suitable for use in modern, factory-made furniture, not for traditional upholstery. More recently, new materials such as sheets of rubberized horse or cheaper hog's hair and ready-made rolls of padding for seat edges have been produced.

*A**lthough modern labor-saving materials are available (above), many upholsterers still prefer to use traditional techniques.*

*T**he most dramatic change in the appearance of upholstery (above right) came with the introduction of the coiled spring and deep-buttoning.*

*A**smooth, plump look is produced today (right) with cheap and labor-saving polyurethane foam.*

The fabrics used for the top coverings on upholstery have varied considerably over time according to availability, cost, fashion and technology. The names of the fabrics have varied too, and it is not always possible to determine exactly what they were from some references. Some of the materials have charming, almost romantic names such as cheney, harateen, paragon, tabby, tinsell and shalloon. The most luxurious have remained constant, such as silk and velvet, both of which can be plain or patterned, or brocaded with gold and silver threads.

Until the Spitalfields silk industry was established in London shortly after 1700, the most fashionable silks were made in France and Italy, notably in Lyons, Tours and Genoa. Genoa was particularly famous for its silk velvet, which was a sumptuous, large-patterned fabric with a silk pile on a plain or satin ground. Genoa velvet was especially popular between 1670 and 1750 and it was widely imitated in both France and England. The Spitalfields silk makers, who produced fabrics to meet both domestic and foreign demand, equalled their continental rivals in terms of quality, but relied totally on France for their designs, from where virtually all fashions in the decorative arts emanated.

Not all velvets were silk; some, such as moquette, caffoy and plush, had a woolen pile and were far more robust. Large quantities of mostly plain-patterned velvets were made in Holland and are now often mistakenly called Utrecht velvets; there are no records of any velvet being made in Utrecht at all.

Other woolen or worsted materials were extensively used for upholstery for furniture in daily use, though not on furniture which was just for show. Many of these materials were plain, but some, such as serge and camlet, had a ribbed pattern woven into them. Other patterns were achieved (as they were on silk as well) by watering, waving and figuring, all processes which changed the nature of various areas of the fabric so that light was reflected in different directions. Because woolen fabrics were so vulnerable to moths, their use has been drastically underestimated.

The most consistently popular upholstery material has been damask, a single-colored

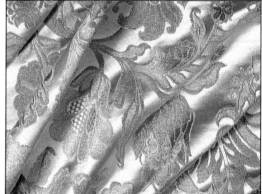

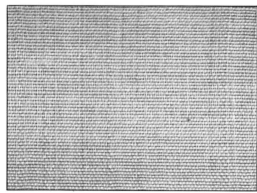

*B*rocaded silk such as this Italian damask (above left) was the most luxurious fabric in the seventeenth century.

*E*arly furnishing silks mostly had large formal patterns such as that on this seventeenth-century French brocatelle (top).

*W*oven horsehair (left) is now supplied in previously unobtainable colors.

*M*any reproduction fabrics have been successfully copied from early dress silks (above).

16

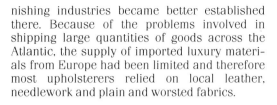

nishing industries became better established there. Because of the problems involved in shipping large quantities of goods across the Atlantic, the supply of imported luxury materials from Europe had been limited and therefore most upholsterers relied on local leather, needlework and plain and worsted fabrics.

DECORATIVE TRIMMINGS ·

Until about 1900, elaborate trimmings were a prominent feature of upholstery and were sometimes more decorative than the fabric they bordered. The gold and silver fringes and brilliantly colored silk tufts and tassels made in the seventeenth century might look out of place in today's homes, but the deep, tasselled, latticework fringes which disguised Victorian chair legs can make all the difference to their appearance now.

cloth, worsted or cotton, in which pattern is achieved by the reflection of light on two contrasting weaves. Early damasks had large baroque patterns, many of which were made in Genoa. Damasks are still popular today.

Turkeywork was a European woven imitation of imported Turkish · carpets. These carpets were highly prized possessions in the late sixteenth and seventeenth centuries. The fabric was extremely hard wearing and brightly colored, its designs based at first on Oriental models, but after 1625 on contemporary needlework.

Painted and printed cotton chintzes were first imported from India in the seventeenth century and were so popular that they threatened the English woolen trade. When their importation was banned by Act of Parliament in 1701, English textile printers began to produce passable imitations until they too were prohibited in 1720. Printers got around the legislation by printing the designs, not painting them, on a cotton and linen mix fabric, fustian, until the repeal of the Act in 1774. Particularly high-quality chintzes were produced in France. They were called *toiles de Jouy* after the town where the largest number were made.

In America the choice of fabrics followed a slightly different pattern until the second half of the eighteenth century, when the various fur-

*E*ven when they are authentic, the use of elaborate tassels and fringes can look a bit too fussy for modern taste (above).

*A*s the most enduringly popular traditional upholstery fabric, damask (right) is now available in almost any color.

*F*lame stitch, sometimes called bargello, point d'Hongrie or Irish stitch, can now be produced on a loom (left).

*U*ntil recently, wool fringes suitable for early furniture (above) have been in scarce supply.

RESTORATION

ntil the nineteenth century upholstery was an expensive and highly prized possession and was consequently supplied with a protective fitted cover. These case covers were sometimes leather, but mostly made from cotton or linen woven with a broad check. They were secured to the furniture with ties or metal hooks and eyes. Case covers appear frequently in eighteenth-century paintings and it seems that it was acceptable to leave them on when a room was being used informally by the family, but they were removed for more formal occasions. Towards the end of the century they appeared with printed designs and began to take on the role of the more permanent and decorative loose cover. In the nineteenth century light-colored floral chintz loose covers were very common and were often used for drawing-room furniture during the summer. They were sometimes removed for the winter, when the somber-colored fixed upholstery was considered more appropriate.

In France, and to a lesser extent elsewhere, the elaborate top covers of some seventeenth-century furniture were also loose and were removed and stored away while the room was not in use. It was not uncommon for grand houses to have two complete sets of upholstery (wall-hangings and curtains as well as seat covers and bed-hangings), one for summer and one for winter. Some houses were even equipped with four sets, one for each season. While this total change-over may seem extreme to us now, it helps to underline the degree of importance which was attached to upholstery in the past.

The daily use which upholstered furniture now receives means that it frequently needs re-covering, and perhaps re-stuffing. Fortunately there are plenty of traditional upholsterers around who appreciate the importance of getting the shape and fabric right. Textile manufacturers have responded to recent demands for traditional materials and offer a wide range of furnishing fabrics with designs based on old documents. ("Documents" in textile jargon meaning both designs and actual surviving samples.) It is worth noting though that many of these fabrics, especially those based on eighteenth-century printed designs with *small* patterns, were originally intended for dresses.

While there are a number of specialist weaving companies who still work on old looms and will undertake individual commissions for exact copies of historic fabrics, the majority of reproduction textiles are not precise representations of originals. This is partly due to contemporary taste, but mostly to practical considerations. Looms are now much wider than previously — early looms could produce fabric

As the prices of antique furniture rise, so the quality of reproductions improves. This fine repro chair (left) is sympathetically covered in silk damask.

A much broader stripe than would originally have been used for this early Victorian sofa has been chosen to counter-balance the pronounced curves (below).

up to a maximum width of only 54 centimeters (21 inches) — and many patterns have to be slightly adjusted to fit them. Synthetic dyes, which have the advantage of allowing a greater number of colorways, often have a harsher quality than vegetable dyes, and in the same way, perfect mechanical repeats lack the depth and character of hand-printed patterns. It can be said, though, that modern textiles are a great improvement. Broader widths mean fewer joins; more varied colors allow a greater freedom of choice; and modern technology has produced materials that are generally cheaper, more durable and far easier to clean.

The greatest advantage of the present situation is the wide range of choice on offer. For the purist, who wants to re-create an authentic, historical interior, there are both craftsmen and materials available. For the majority, who like the traditional look but still want their homes to remain in the twentieth century, it is possible to capture the mood of a particular period in history with fabrics in the right style, but not directly copied from earlier examples.

19

It is perfectly possible to remove and restore the original covering of a valuable piece of furniture such as this Art Deco sofa (top) by Paul Follot.

The bolder colors, made possible with modern dyes, produce dramatic effects, particularly when applied to traditional designs (above).

Because the shape of the upholstery on this reproduction chair (right) is correct, its modern cover and border don't look out of place.

THE STYLES OF THE COURTS

LUXURY AND GRANDEUR
·
DUTCH AND
FLEMISH INFLUENCE
·
A RETURN TO LUXURY
·
QUEEN ANNE FURNITURE
·
THE NEW WORLD

Caned Restoration chairs
at Ham House

LUXURY AND GRANDEUR

ntil Louis XIV assumed power in France in 1661 the decorative arts in Europe had been dominated by the Italian Renaissance and baroque design. A great deal of furniture was imported to France from Italy, Germany and the Low Countries and immigrant Italian and Flemish masters were employed in the Louvre in Paris to instruct French craftsmen in the arts of furniture making. By the middle of the century, French furniture was increasingly characterized by opulence — ebony and other veneers, and inlays of precious materials appeared on cabinets and armoires, while tables and seat furniture had elaborate turned and carved supports. Comfort had become an important consideration and fixed upholstery was common. The square-framed, low-backed chairs were gradually replaced by the *fauteuil*, a broad, high-backed chair with curving carved arms. The back tilted slightly backwards to allow the sitter to recline.

· FAVORITE FURNITURE ·

Upholstered furniture was generally made in large suites and now included a variety of seating in line with the status of a room's occupants. The most important people sat on chairs with arms; the next in line on those without; women usually sat on *tabourets*, which were often of X-frame construction, either folding with a loose cushion, or solid with fixed upholstery.

Most suites included at least a pair of *porte carreaux*, very low stools on which one or more large, soft cushions were placed. Cushions of this type were rather floppy and usually had tassels attached to the corners which were themselves drawn in towards the center of the cushion, forming an indentation from which the tassels protruded. In Spain it was customary for ladies to sit on a squab placed directly on the floor, the underneath being made of leather. Cane and rush-seated chairs, of rustic type, were also common, even in grand houses, and they too were supplied with a cushion.

In the bechamber, apart from the elaborately draped and canopied bed, it was customary to have a *lit de repos*, a single-ended upholstered daybed. These might be fitted with a loose squab mattress and a bolster, or a large cushion, over an upholstered seat and back. In grand houses there may also have been a doubled-ended, backless *couche* which was placed against the wall under-

In this fashionable French interior of 1635 by Abraham Bosse (above) the fabrics and trimmings of the bed, table carpet, folding stools and richly upholstered chairs are coordinated.

Design Detail

Porte-carreaux were low stools onto which these large, soft cushions were placed.

This reproduction fauteuil *(right) is correctly covered in rich silk and properly shaped with a domed seat and thin, flat back.*

Elaborate multi-colored silk fringes and braids were a simple way of creating a luxurious effect. This chaise en confessional (left), for example, looks quite grand.

The Italian silk fringe (above) would have transformed the appearance of any piece covered in the simple contemporary green damask (below).

neath a canopy. This meant that they were seats of state and not for normal domestic use. The tilted ends supported large cushions and there was often a gathered skirt hanging over, or from, the front seat rail underneath the padded seat.

After 1650 furniture became grander in appearance. When Louis XIV came to the throne, he determined to establish a national style in the decorative arts and to provide a symbol of the greatness of France and her king by glorifying the Palace of Versailles. In 1662 his minister for the arts, Jean-Baptiste Colbert, set up a royal factory, *Manufacture Royale des Meubles de la Couronne*, at Gobelins – and in the following year appointed the artist Charles le Brun as its director and principal designer. Le Brun's greatest skill was as a design co-ordinator, bringing together various foreign influences into one distinct cohesive French style. Monumental baroque splendor was tempered by fine classical details. Cabinets, tables and commodes (chests of drawers) were decorated in imitation of oriental lacquer or inlaid with marquetry veneer of colored woods and ivory, or a combination of tortoiseshell, brass and pewter. Highly carved seat furniture was often painted in bright red or green and sometimes silvered or

gilded. The legs and stretchers were scrolled and the arms of *fauteuils* curved and padded on top.

Around 1675 a new type of armchair evolved known as *en confessional*, on which two wings, fixed at right angles to the back, hid the sitter's face from view. This was a forerunner of the fully upholstered wing chair which was particularly popular in Britain and America. At the end of the century, references to *canapés* or *sophas* began to appear. These were what we call a sofa today, a seat with a back and arms, designed for two or three people and supported on a wooden underframe. Illustrations show that these were fully upholstered and sometimes had loose cushions or bolsters and a skirt running along the front and sides. Before 1700 *sophas* were not common and surviving examples are rare. By 1700 both the backs and seats of *fauteuils* were lower and broader and stretchers were beginning to disappear.

· MATERIALS ·

Upholstery textiles used on Louis XIV furniture include silk, velvet, damask, brocade and needlework, all bordered by elaborate silk or woolen tufted fringes. Sometimes the materials were additionally embroidered. Tapestry was a popular

covering for squab cushions and was often woven specially for the purpose; it was only occasionally used for fixed covers at this date.

Stamped and gilded, or silvered, leather was also used and was edged with widely spaced gilt-headed nails placed over a simple galloon, or braid, or the heading of a fringe. Sometimes the upholstery was paned, or panelled, using two different fabrics, or the same fabric in two contrasting colors. Fringes too were multicolored and the grandest were of gold or silver. Joins in the fabric were often disguised by a band of galloon. Because the method of stuffing chairs was still rather basic, the upholstery tended to have thin edges and a domed appearance.

The influence of Louis XIV furniture spread throughout Europe, particularly after the revocation of the Edict of Nantes in 1685 which forced large numbers of Huguenot craftsmen to flee the country to avoid religious persecution. Despite the departure of some of France's leading designers and a simultaneous loss of love for ostentation on the King's part, luxury and grandeur continued to be the principal characteristics of French furniture for a further 100 years.

DUTCH AND FLEMISH INFLUENCE

ecause Dutch and Flemish furniture was seldom innovative during the eighteenth and nineteenth centuries, it is often forgotten that in the seventeenth century the Netherlands was an important maritime power and consequently a flourishing center for commerce and the arts. It was from the Netherlands that the techniques of veneering and marquetry were learnt by the French and from where the use of woven split cane for the backs and seats of chairs was first copied by the English. Woven cane was discovered by the Dutch in China through the trading activi-

ties of the Dutch East India Company. A passion for Eastern goods spread throughout the Continent. Attempts by European craftsmen to imitate oriental design resulted in a style which has been labelled chinoiserie.

Apart from a fashion for Indian printed cotton palampores (bedcovers), imported oriental textiles had little direct effect on contemporary upholstery in the Netherlands. Dutch and Flemish seat furniture was strongly influenced by that in France and numerous illustrations of interiors show the same long rows of low-backed chairs and carved *fauteuils* covered in silk or velvet and bordered with fringes. The

The Dutch were first to popularize drop-in seats around 1700. This slightly later example (above) has a silk cover with a chinoiserie pattern of pagodas and oriental trees and flowers.

Design Detail

Some of Daniel Marot's designs had covers that draped over the edge like a skirt.

Throughout the seventeenth century interchange of ideas between European countries was considerable. This Flemish chair (right) with its X-plan stretchers is a typical French type.

principal differences lay in the wooden underframe which had the same H- or X-plan stretchers, but was usually turned rather than carved. Barley-sugar twists are characteristic of Netherlandish chairs of the period. Following the introduction of woven cane in the middle of the seventeenth century, large numbers of totally wooden chairs with a mixture of carving and turning were made with caned seats and backs and a broad carved and pierced border set an inch or so below the front seat rail. A simple loose cushion with fringed edges and tassels at the corners was placed on the seat,

The published designs of the French émigré Daniel Marot for magnificent curtains and upholstery in the latest Parisian styles (above) were greatly admired in the Dutch and English courts where he was employed by William of Orange.

Dutch traders were first to import Indian cottons and oriental silks to Europe. Their fruit and flower patterns (right) have been an inspiration to European and American textile designers ever since.

trimmings, particularly in the bedchamber. His designs for chairs are in a variety of types, although all have tall, narrow backs. A few have caned seats with squab cushions and carved and pierced backs, but the majority have fixed upholstery with the seat cover either extending over the edge to form a draped skirt, or directly secured to the edge of the frame with gilt-headed nails. Some have square-edged upholstery on a back and seat of drop-in type, that is, it must have been attached to separate frames which then slotted into the chair frame itself. Some of the designs show matching stools. The neater edges and the flatter seats imply a greater degree of sophistication in the stuffing underneath.

Despite the popularity of Marot's designs with the Dutch court, their opulence did not appeal to the mostly Protestant and anti-aristocratic middle-class merchants who constituted the largest and most prosperous section of Dutch society. The lavish upholstery made for the royal residences was therefore seldom seen in other houses. More moderate chairs, usually combining a tall, carved and pierced, or carved and caned, back with a fully upholstered seat were made in large numbers. Leather and various woolen and worsted fabrics were more common than silks, velvets and brocades.

but the cane on the back was usually left exposed. This distinctive type of chair was also popular in Restoration England, but was seldom seen in France and elsewhere.

· DANIEL MAROT ·

The most influential Huguenot designer to leave France in 1685 was Daniel Marot, who was immediately employed by William, Prince of Orange, later William III of England, to redecorate his palace at Het Loo. Marot subsequently followed William to England.

During the 1690s he published a series of engraved designs for furniture and interiors in the full-blown Louis XIV style and they show a strong emphasis on luxurious upholstery and

RE-COVERING A DROP-IN SEAT

Regardless of the shape of the seat, cut your fabric in a square, allowing a bit extra for folding over at the sides. Mark the center of each side with chalk and do the same to the frame, making sure that any pattern will be centered on the seat. Lining up the chalk marks, attach the fabric in the middle of each side with a 12mm (½in) fine tack tapped in just halfway to allow for repositioning.

Working away from the center, continue along each side, smoothing the fabric evenly over the edge as you go. At the corners, fold the material to make pleats on the front and back, making sure that the open edges are on the outer sides of the frame. When all the tacks are in place, hammer them home, cut away any excess fabric and fix some bottoming (upholsterers' linen) across the underframe to finish it off, folding the edges under neatly.

A RETURN TO LUXURY

At Knole, a mansion in Kent, there is an example of the more comfortable furniture that was being seen in England before the civil war. It is a style still seen today — a fully upholstered couch with hinged headrests at either end, a sort of double-ended daybed. The headrests on these daybeds were adjusted by ratchets.

British furniture design then made little progress under Oliver Cromwell's Puritan rule. The carved Renaissance motifs which had previously decorated rather ponderous oak furniture were replaced by simpler moldings and turnings. For seat furniture, various stools and the square-framed farthingale chair remained the most common types alongside a much smaller number of slightly wider chairs with either padded or wooden arms. Leather upholstery, held in place with large-headed nails, replaced the brightly colored turkeywork and other woolen fabrics which had been so fashionable before.

When Charles II and his court returned from exile they brought with them a desire for the luxurious furnishings they had seen abroad. The King determined to establish in England the high standards of design and craftsmanship he had encountered in France and Holland. A reaction against Puritanism, and rising national prosperity resulting from successful colonization and trading in distant parts of the world, combined to produce a situation in which the aristocracy, the landed gentry and the increasingly wealthy merchant middle class were receptive to new ideas and were eager to acquire comfort and status by the acquisition of rich furnishings. The furnishing industry also received an unexpected boost when the Great

Design Detail

The elaborate dolphin chairs at Ham House have sumptuous silk upholstery fixed with hooks and eyes.

Knole House in Kent contains some of the best surviving original upholstery in the country. The modern Knole sofa is based on an early seventeenth-century example in the house. The Cartoon Gallery contains an assortment of Restoration and Queen Anne pieces.

26

This anonymous, late seventeenth-century painting of a bedchamber (left) shows marvelous detail. The matching hangings and upholstery have long gold fringes and the chairs are bordered with braid and gilt nails. The backs of the chairs, which only the servants saw, are covered with plain, coarse cloth.

While leather had been fashionable for upholstery under the Commonwealth, it was less common on Restoration chairs. Plain, nailed borders were always favored for leather, but chinoiserie decoration is most unusual. Similar-shaped seat edges to these (below) were seen on Marot designs.

Fire of London destroyed large numbers of houses and their contents in 1666. Because London was the home of the court, around which all fashionable people gathered, it also became the center of the furnishing industry and upholsterers and decorators were traditionally based there.

The most common type of seat furniture to appear after the Great Fire was the carved and turned chair with caned seat and back which had evolved in Holland. Interestingly these chairs seem to have been popular in small houses long before they appeared in the grand homes. Originally they were quite small and had simple spiral turning all over, but they soon became more elaborate, with taller backs and more decorative combinations of open carved and turned areas. Single-ended daybeds, based on French *lits de repos,* were also made of this type, and like the chairs were supplied with loose cushions, but of more substantial mattress type with a border. Cane chairs were so popular in England that the Guild of Upholster-

ers felt moved to petition Parliament (unsuccessfully) to prevent the importation of canes. Despite the almost universal popularity of cane, fully upholstered furniture in the French style was favored for the most important rooms of a grand house.

· HAM HOUSE ·

The arrival of Huguenot craftsmen in London after 1685 and the published designs of Daniel Marot undoubtedly encouraged the production of elaborate upholstered furniture. It had, however, already been produced in large quantities before they arrived. One of the most impressive houses of its day, Ham House in Surrey, was extravagantly refurbished during the 1670s in the latest taste and most of its furniture and a great deal of its original decoration have survived intact. Missing upholstery has recently been carefully reproduced according to the original inventory descriptions and gives a good idea of the brilliant colors and textures.

Most of the family rooms at Ham were hung

with silk damask, the colors including crimson, black, white, gold, yellow and blue. The state rooms upstairs had grander wall hangings of silk, satin and velvet, some of them embroidered or brocaded in gold or silver. The upholstered furniture was fashionably covered to match. Gold fringes were used for the grander pieces and silk tufted fringes, mostly in two colors, were used downstairs. The majority of the silk fabrics must have been imported from France or Italy, while those with a wool base were probably made in England or Holland.

Many of the chairs at Ham have fashionable gilt frames; the remainder are made of walnut, or painted, japanned or grained beech. When marquetry and other decorative veneers of highly figured woods, such as yew, maple, olive or laburnum, became fashionable for cabinet furniture and tables, oak was no longer considered suitable for chair frames. Walnut became the most common and most admired substitute for both chair and table legs.

· DUTCH INFLUENCE ·

After about 1680 greater numbers of arm, or elbow, chairs were made. Some sets of chairs had equal numbers of elbow chairs and backstools which were arranged alternately around the room giving the appearance, at first glance, that they were all armchairs. In line with those in France and Holland, the backs of chairs became slightly taller.

While *sophas* were evidently a great rarity in England before 1700, upholstered daybeds were quite common in bedchambers. Upholstered stools of the period seem to have been rarer than in France, possibly because the rules of etiquette were less rigid and chairs or backstools were acceptably used by greater numbers of people.

During the 1680s England saw the emergence of the wing chair. Early references describe them as *easie* chairs, which implies that they were intended for relaxation, not for formal use. They were to be found in closets, the small private rooms or studies which in grand houses were placed beyond the bedchamber, where they were inaccessible to all but a favored few. This proves that they were for personal use only.

The sumptuous and colorful interiors of Restoration England were magnificent compared with earlier times, but paled in comparison with the splendor and grandeur of Louis XIV's royal palaces. Like the Dutch, the English admired and acknowledged the French as arbiters of taste, but their more conservative nature could not allow them the same extravagances. In view of the succession of William

*T*he Genoa velvet, which still covers this magnificent 1690s daybed (above), was particularly fashionable around the turn of the century. Red and green on a cream background was the most common color combination.

*A*lthough a little the worse for wear, the original tufted and tasselled fringes and decorative borders of galloon are in place on this early example of an English sofa (left). The down-filled cushions are still plump.

28

*T*he wings of this 1670s
adjustable-back sleeping
chair at Ham House (left)
protected its occupant from
drafts. The gilded frame,
silk covers and gold fringes
were considered very
grand in England. To the
right is a large, floppy
velvet cushion with
tasselled corners, recently
reproduced to fit one of a
pair of low squab frames.

of Orange to the English throne in 1688, it is
hardly surprising that there were for a while
closer links between England and the Netherlands than between England and France and
English furniture design more closely followed
Dutch models.

Like the Dutch, the British had formed an
East India Company at the beginning of the
century and traded successfully in the Orient
and the East Indies. By 1680 Indian cotton or
callicoe palampores had become a popular
form of bedcover and before long imported
cotton was being used for window curtains,
cushions and chairs. Because it was in short
supply and expensive, there were many attempts made to copy it, on the one hand by
textile printers and on the other by Restoration
ladies. These women worked exotic and bold
naturalistic patterns centered around their
favorite motif, the tree of life, with crewel — a
fine-spun worsted embroidery thread — on a
linen or twill ground. Few examples of crewel
work have survived other than bed-hangings,
but at the time it was felt to be an ideal fabric
for use in conjunction with imported oriental
lacquer furniture or European japanned
versions.

QUEEN ANNE FURNITURE

The first two decades of the eighteenth century saw furniture design in England follow two separate lines of development. That produced for the royal palaces and the large town and country houses of the aristocracy and political magnates was still influenced by French design. Richly upholstered suites of chairs, stools and magnificently draped beds of enormous size continued to be regarded as symbols of wealth and power. Gilded furniture was particularly fashionable and ornate carving continued to be a prominent feature on all furniture.

On the other hand, the landed gentry and the respectable merchant class, who had neither the houses nor the money to support such ostentation, purchased elegant, well-made furniture with little ornamentation. This category of furniture is chiefly characterized by plain, smooth surfaces and the dominant use of walnut veneer, for seat as well as cabinet furniture and tables. Its most novel feature was the cabriole leg. Because the addition of stretchers to the cabriole legs interrupted the flowing line, they were soon dispensed with. The broad knee of the leg was fashionably carved with a shell and the feet were usually in the form of a hoof or pad. This distinctive shape originated in China where it was presumably noted by Dutch and English East India Company employees. It has been suggested that it was simply a logical extension of the baroque S-scroll, a double version of which frequently formed the legs of late seventeenth-century chairs. This type of walnut domestic furniture is invariably described as Queen Anne (1702-14), although in reality most of it was made during the reign of her successor, George I (1714-1727).

The appearance of so-called Queen Anne upholstered furniture is totally unlike its grander contemporaries. Armless or single chairs were of two types — either fully upholstered, or wooden with a drop-in seat. The former were usually square-seated with a straight-sided back which was sometimes rounded at the top.

*E*mbroidered crewel work with patterns based on Indian cottons continued to be fashionable during the reign of Queen Anne. The tree of life was still the most commonly reproduced motif (above).

Design Detail

Drop-in seats were frequently covered with a needlework design.

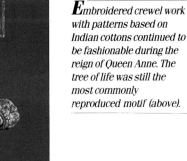

*S*ilk damask remained the most popular covering for drawing-room furniture throughout the eighteenth century. This Kentian sofa (left), with its reproduction upholstery and nailed borders, has been given a pair of tassel-cornered cushions, a feature which is often missing, but which can make all the difference to the final appearance.

Needlework was found on Queen Anne chairs of all types and was often worked by the lady of the house. Drop-in seats (left) became the norm for side chairs.

This unusually wide chair (above) with its faded Genoa velvet upholstery is one of a set at Ham House dating from about 1730. Its rather thick, plain, crimson piping is an early example of an increasingly popular form of edging.

Tapestry does not seem to have been used much for upholstery in England; this example (right) was made at the Soho Works. The traditional gap between the back and the seat is filled by this date.

31

The back of the frame was covered over by the upholstery. The stuffing was no longer domed but was uniformly flat, with smooth, slightly rounded, shaped edges.

Wooden-framed chairs were more curvaceous. The seats were rounded in shape and had a deep rail, which was rebated to take the drop-in frame. The backs were molded to support the back of the sitter more comfortably and in section were of swan's-neck shape, usually having a serpentine top-rail and a broad, vase-shaped splat. On armchair versions, smooth, plain arms of a twisted S-shape continuing down into a scroll — now popularly called a shepherd's crook arm — were fixed several inches back from the edge of the seat. After about 1720 armchairs with wooden arms, but fully upholstered seats and backs — sometimes referred to as spoonbacks now — were very popular. However, the most popular chair was the fully upholstered wing chair. Very early examples sometimes had rectangular wings, but by 1700 they were rounded. The arms of the chair rolled outwards and ended in a prominent scroll several inches back from the front of the seat. There was a clear join between the wings and the arms. After about 1710 cabriole legs replaced the earlier baroque underframe. The upholstery was even and quite thick and the seat has a fat down-filled cushion, a feature copied from the French.

Sofas, which were effectively composed of two wing chairs joined together, were a rarity in England before 1700. A new type of seat furniture evolved from the old-fashioned wooden settle, the settee. The word settee has been used in this century to describe the modern sofa, but correctly refers to a long seat with wooden arms and usually a wooden back. The upholstery of the seat either extended over the seat rails or was attached to a drop-in frame.

Comfortable sofas and wing chairs seem to have replaced the daybed (which was still made but in much smaller numbers), presumably because they could be used in drawing rooms as well as bedchambers and in smaller houses took up less room.

INFLUENCE OF FASHION

As so often in the design of furniture, a change in fashionable dress altered the shapes of some chairs. While ruffs were worn, the chair backs, for example, were low so as not to push the wearer forward. When wigs extended to amazing heights at the end of the seventeenth century, so did the backs of the chairs, sometimes more than twice as high as the distance from the seat to the ground. The undergarment known as the farthingale had been abandoned earlier in the century and the gap between the seat and the back was closed in for the first time in Queen Anne styles. Later the crinoline required the arms of the chair to be set further back.

The flourishing silk-weaving industry in Spital-fields provided the silks and velvets seen on grander chairs of the period. This English manufacture was worked to a large extent by immigrant craftsmen. Another valuable material used, though only rarely, was tapestry. There had been a flourishing tapestry manufacture at Mortlake on the River Thames, but in the early eighteenth century the principal factory was in Soho. Wall hangings formed the major part of the output, but specially designed chair covers were also produced.

The most popular form of covering for domestic furniture was needlework. Although ready-made covers could be purchased from professional needlewomen, it was common for the ladies of a household to work their own. The designs were mostly floral, but sometimes the back contained a small picture, usually copied from an engraving. The presence in some pieces of exotic birds and the tree of life motif is evidence of the influence of imported Indian cottons, but presumably printed patterns from other sources were also copied. Geometric patterns were popular too, in particular Irish stitch (*point d'Hongrie* in France), a pattern of zigzags of uneven height.

Another pastime for these women was knott-ing. The pattern was drawn out on a piece of material and then lengths of knotted thread, produced on a small shuttle, were stitched in place over the lines. It was also common to make the tassels and fringes although the majority were made by professionals. After about 1710 fancy trimmings were mostly confined to bed-hangings and curtains and, apart from simple braids, were seldom used on seat furniture. Widely spaced gilt nails frequently held the braid in place.

· PALLADIANISM ·

After about 1725 the influence of Palladianism was felt in English furniture. The chief exponent

*P*ine or oak panelling with classical carving was fashionable in Queen Anne interiors. Although the early wing or easy chair, with its fashionable needlework (right), is contemporary with the room, the chair on the left is some 20 years earlier and has turkeywork upholstery. As today, most homes had an eclectic assortment of old and new furniture.

*I*t is hard to believe this unused satin chair cover with knotted decoration (above) is over 250 years old. Its brilliance is almost garish.

*T*he State Bedchamber at Powys Castle in Wales (left) contained the grandest furnishings. The silvered furniture is covered with Spitalfields silk velvet.

was the architect William Kent, who had traveled to Italy and been greatly impressed by the symmetry and proportion of classical architecture and in particular by the Renaissance interpretation of it by Andrea Palladio. At the same time the destruction of the walnut crop in Europe following severe winter frosts had produced a shortage of timber, and mahogany, which had been available for some time, suddenly became the favorite wood of both cabinet and chair makers. Because mahogany carves so well, carving became a prominent feature of furniture of all types.

In general chairs became broader and cabriole legs thicker and heavier. Usually they ended in claw-and-ball feet. Arms were often raked back at the top further than at the bottom. Red and black japanning was a popular alternative to mahogany and grand chairs were made with a wooden frame completely covered with carved and gilt gesso, a sort of plaster composition.

Upholstery fabrics changed very little. Needlework remained the most popular form of covering for furniture in domestic use and silk and velvet, particularly that produced in Spitalfields, for state rooms. Leather was quite common, especially for dining chairs.

THE NEW WORLD

By the end of the seventeenth century towns along the east coast of America were well established and the professional people and prosperous traders who made up the wealthiest sector of society were demanding more sophisticated furniture for their new houses than the utilitarian furniture of the early settlers. Furniture making in America was hampered by a shortage of skilled craftsmen and difficulties in obtaining good tools and materials. Between them joiners and turners produced simple furniture based on English Jacobean models, which was brightened up with painted or stained decoration. The most common forms of seating were plain wooden stools and benches, but after about 1650 rush-seated chairs with turned uprights and decorative spindles on the backs and arms were made in large numbers and continued to be popular until about 1720. The most common types were called Brewster or Carver chairs after the dignitaries who had supposedly brought their prototypes with them from England on the *Mayflower*.

Chairs with wooden arms of the Cromwellian type were made with imported turkeywork or leather upholstery on the seat and the back bordered by large decorative nails. The best leather, which had a small, coarse, diamond pattern, was imported via England from Russia, but local hides were mostly used. Woods used included black walnut and oak and, to a lesser extent, pine, maple, ash, hickory, birch and fruitwoods, such as pear, apple or cherry, depending on the locality.

At the turn of the century the most prosperous towns were Philadelphia, Newport and Boston, and it was in the latter that the fashio-

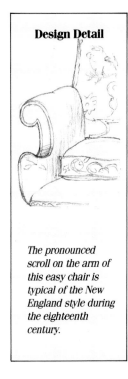

Design Detail

The pronounced scroll on the arm of this easy chair is typical of the New England style during the eighteenth century.

The number of people who could afford expensive foreign furnishings was initially limited, and the majority of chairs were covered with leather, needlework or worsted materials. Check linens were also popular for the loose cushions.

nable furniture of the day was made. Anglo-Dutch carved chairs and daybeds with panels of woven cane in the seats and back were made in sizeable numbers, fitted with cushions of local patterned worsted materials. Large numbers of chairs with simple, turned frames, leather-upholstered seats and tall backs, also covered in leather, were exported from Boston to other parts of the country. Later known as Boston chairs, they were popular until 1770.

· QUEEN ANNE STYLE ·

Around 1720 furniture makers began to adopt the Queen Anne style, which persisted until at least 1760. Elegant but substantial chairs and stools were made, mostly in mahogany and walnut, with broad, vase-shaped splats on the back, cabriole legs joined by stretchers, and wide, flat, drop-in seats. Carved shells and fan motifs were often added to the top of the back, the middle of the front seat rail and the knees of the front legs, although the latter sometimes had carved acanthus leaves. Feet varied from the simple hoof or pad to the more elaborate claw-and-ball variety.

There were plenty of regional differences. New York chairs, for example, had rounded balloon or horseshoe-shaped seats instead of the straight seats made elsewhere. They were also made without stretchers.

The most sophisticated upholstery was seen on the easy chairs found in bedrooms. On examples made around 1710 the arms and the top of the back had a pronounced scroll, but later they were gently curving. New England easy chairs had vertically rather than horizontally scrolled arms. Like other chairs and stools, the majority had cabriole legs joined by turned stretchers, a feature which persisted on easy chairs until as late as 1770. During the 1750s similar English-style sofas were produced in Philadelphia with simple cabriole legs (without stretchers), rolled arms and a serpentine back.

Because of the difficulties in obtaining rich upholstery materials — mostly imported from Europe — chairs of all types were fashionably covered with patterned worsted fabrics or needlework. Some needlework patterns were copies of those seen on more luxurious damasks and velvets, but the most popular were floral patterns and Irish or flame stitch. Crewel work seat covers were sometimes made for side chairs. For similar reasons, until the latter part of the century, few windows were supplied with curtains and wall hangings were not at all common. Most rooms had wooden panelling or walls that were painted, or, later, papered.

35

*A*merican furniture in the first half of the eighteenth century was similar to that made in England, although the Queen Anne style lingered on until about 1760. Luxury fabrics, such as this type of silk damask, were mostly imported from Europe.

*T*here was a marked tendency among American makers to retain stretchers on Queen Anne-style furniture long after they were dispensed with in Europe (left). It took a long time for new fashions to cross the Atlantic.

THE STYLES OF THE COURTS

THE ROCOCO

FRENCH ROCOCO

·

THOMAS CHIPPENDALE

·

AMERICAN CHIPPENDALE

A rococo veilleuse by J.B. Tilliard, 1750

FRENCH ROCOCO

uring the first half of the eighteenth century life in the French court was more informal than previously. Rooms were smaller and more intimate and although the furnishings were as luxurious as they had been under Louis XIV, comfort was now an important consideration.

The smaller size of rooms didn't mean that any less money was spent on them. Encouraged by the *marchands-merciers* — dealers in furniture and decorative objects — the aristocracy paid out enormous sums for fashionable furniture, paintings, fabrics, ornaments and *boiseries* (carved and painted wall decoration) — all of which had to complement each other in style and color. Sometimes the design of the furniture repeated the design and color of the wall decoration against which it stood.

The dominant style during the regency of the Duke of Orléans and the reign of Louis XV was rococo, characterized by continuous flowing curves and asymmetrical arrangements of natural forms such as shells, rocks, and flowers.

To suit the more relaxed social atmosphere, new types of seat furniture developed which were lighter and more comfortable. Seats and backs were lower and the use of the cabriole leg was universal, meaning that stretchers completely disappeared. The backs of chairs were slightly bowed and the upholstery slightly concave to accommodate the back of the sitter more comfortably. After about 1720 the arm supports were set several inches back from the front rail to prevent hooped skirts from being crushed. Upholstered armrests became a standard feature.

Although some spectacular chairs and sofas were designed in full-blown rococo style, a much simpler version evolved which was used by *menuisiers* in the provinces as well as in fashionable Paris. The entire frame was composed of continuous curves, the line of the cabriole legs flowing into and around the seat rail and then up through the arm supports, along the arms and up and along the cartouche-shaped back. Relief decoration in the form of carved flowers, shells and other rococo motifs was added at appropriate intervals.

The frames were made in solid walnut, beech, oak or fruitwood, such as maple or cherry, often left in their natural state and simply waxed and polished. The most expensive were gilded, silvered or alternatively painted – green, pink, blue, yellow or silver – or coated with colored varnishes.

However elaborate the frame was, the upholstery invariably cost considerably more. Velvet and damask were the fashionable coverings for winter use and printed cotton or embroidered or painted silk for summer. Embroidery silks and patterns were available for sale to fashionable ladies who wanted to make their own seat

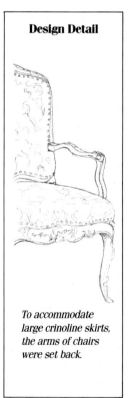

Design Detail

To accommodate large crinoline skirts, the arms of chairs were set back.

T his Regency winged beech canape (left) shows the beginnings of the rococo style in the rather hesitant shape of the cabriole legs and the retention of stretchers. Needlework was a popular covering.

I n France stools remained an essential part of any suite of seat furniture. In its arrangement of C and S scrolls, this traditional X-frame stool (above) has been cleverly adapted to suit the new rococo style.

*T*he French rococo style inevitably spread, but was not always successfully adapted. This mid-century German chair (left) has a skirted seat cover reminiscent of Marot's pre-1700 designs.

*T*he tapestry covers on this typical Louis XV fauteuil (above) are later replacements, but give a good indication, with their close-nailed borders, of the original appearance.

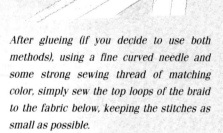

covers. Because it was too expensive for most people to have more than one complete set of furniture, and the covers were frequently changed, and because the wood frame was totally exposed, various methods of attaching the upholstery with screws and clamps were devised, although these were by no means common. Upholstery was normally fixed with closely spaced gilt nails.

· VARIETIES OF FURNITURE ·

There were many different types of chairs and sofas, usually designed to match the rest of the room. *Sièges meublants* (consisting of *fauteuils* and sofas) stood permanently against the wall; *sièges courants* were pieces intended for use in the middle of the room, for activities such as conversation, sewing and card-playing; straight-backed side chairs for use at a table were known as *à la reine*, those with a concave back *en cabriolet*. *Fauteuils en confessional* became more popular, but not as popular as *bergères* — large, low armchairs in which the space under the arms was also upholstered. The seat was given a thick,

down-filled loose cushion for extra comfort. Because card-playing was such a popular pastime, a special type of chair evolved called a *bergère voyeuse*, which had a flat-padded top on which onlookers could comfortably lean over the shoulders of those playing.

Sofas too came in many different forms. As well as the standard *canapé*, with its upholstered back and arms, there was the *confident* (or *tête-à-tête*), a deeper sort of *bergère*, the *ottomane*, with its oval seat, and the *marquise*, which had a straight seat. The *duchesse*, really a development of the *lit de repos*, was made up of either one or two *bergères* and a close-fitting, large stool. Suites of upholstery still included a number of *tabourets* and matching fabric-covered fire and folding screens as well.

Many surviving Louis XV chairs have tapestry covers. As tapestries were still widely produced in the eighteenth century and a number of documented sets of eighteenth-century French tapestry covers are known in England, for a long time it was assumed that they were all original. They are in fact mostly nineteenth century.

THOMAS CHIPPENDALE

n view of the eighteenth-century Englishman's admiration of the French as leaders of fashion, it seems surprising that the rococo style was not adopted in Britain until nearly the middle of the century. Even then the style was really considered too excessive for English taste and most people were quite content with furniture simply described as "in the French taste."

The most successful interpreter of the rococo style was the English furniture maker Thomas Chippendale (1718-79), whose published designs for furniture and upholstery of all types were widely copied throughout Britain and in America. Not all his designs were French in style; many were gothic and chinoiserie, and later, in the neo-classical style. Although it was his intention to make furniture "suited to the fancy and circumstances of persons in all degrees of life," many of his clients were members of the aristocracy and some of his output was elaborate. The elegant, straightforward mahogany furniture made between 1750 and 1770, now commonly referred to as Chippendale, represents a simplified version of his designs and is very different from the French furniture that largely inspired it.

A large proportion of the seat furniture in his *The Gentleman and Cabinet-Maker's Director* (1754) is in the rococo style and many of the chairs are labelled as French. A book of designs published in the same year by another large firm of cabinet-makers and upholsterers, Ince & Mayhew, included *French Burgairs, French Stools* (window stools), sofas, *French Corner Chairs*. Single chairs, with delicately carved and pierced splats, are all described as *Back Stools*. In both books rococo carving on the chair designs is detailed, more so than the carving on most French chairs of the time.

Chippendale suggested that chairs should be covered in Spanish leather, damask, tapestry or needlework, although the latter two were rapidly becoming unfashionable. Some of the chairs show lightly sketched tapestry designs on the seat and back. In reality the most popular material for covering drawing-room furniture was damask; crimson was the fashionable color and the most expensive because it was so difficult to dye. Chinese painted silk was sometimes used for chinoiserie furniture. Leather and a new fabric, horsehair or "haircloth," which was woven with actual horsehair,

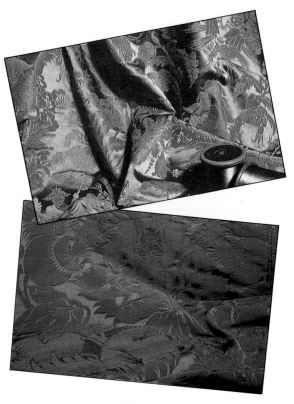

40

Design Detail

Silk tufts were the precursors of shallow buttons.

*S*ilk damask was the most popular upholstery fabric for Chippendale-style furniture and is available today in a vast variety of different colors and patterns (above).

*T*his rather strange design for a window stool by Ince & Mayhew (left) has fashionable shallow-buttoning of silk tufts, as well as a border of brass nails. The tufts would have matched the fabric in color.

UPHOLSTERY STYLES

In his Director, *Thomas Chippendale sometimes sketched in a possible upholstery fabric, such as Chinese painted silk (above).*

This splendid armchair (right) appears in a 1770s portrait by the painter Richard Cosway. It has been re-upholstered with the same crimson damask and matching silk tufts. The padding on the arms is very domed, more so than modern upholsterers would expect.

*H*ow not to upholster a Chippendale armchair (above) — the bland color and flat upholstery of this otherwise vigorous transitional piece is totally out of keeping. Note how both the back and the seat are set in.

was popular for libraries and in particular for dining rooms because it could be wiped clean and didn't retain cooking smells. At this date horsehair could be supplied in black, red or green, and was either plain, striped or checked.

Very occasionally the edges of the fabric were concealed by a narrow strip of brass, but the most common form of trimming was still a single or double row of gilt or brass nails. Ince & Mayhew showed a number of parlor (single) chairs, with ornamental patterns for nailing, but they don't seem to have been copied in practice.

By 1750 the stuffing of the upholstery had become quite sophisticated and was able to follow the contours of the frame. The shape of

the padding varied according to the style of the piece. French chairs in the rococo style had uniformly flat, round-edged seats, usually bordered by a single row of nails. The stuffing of the back was often thicker and well rounded and was sometimes held in place by several rows of silk tufts, creating a shallow-buttoned effect. The padding on the armrests was domed.

On chairs in the Chinese and gothic styles, which had straight "Marlboro" legs joined by stretchers, the seats were often of box shape, the top edges where the seat fabric joined its border being defined by piping. Double rows of brass nails running around the top and bottom of the seat rail were the most common form of border. If the straight back of the chair was up-

*T*he simple chair (above) has a fully upholstered back and brass-nailed seat. Apart from the completely flat armrests, the upholstery is thicker than the piece opposite, but still lacks depth. A double row of nails would give it more force.

holstered, it too had borders of nails, as did the flat padding on the armrests. Chippendale indicated that his Chinese chairs should have "cane bottoms, with loose cushions, but, if required, may have stuffed Seats and Brass Nails." The seats of straight-legged chairs were either straight or serpentine-shaped at the front. Sometimes the seats were dished, the straight sides being higher than the concave front and back. Drop-in seats were still being made.

The upholstery of wing chairs, sofas and daybeds followed the same principles. The back of sofas, whether they had straight or cabriole legs, were serpentine and the arms scrolled outwards. Loose cushions in matching material leant against both the arms and along the back, their number depending on the number of people it was designed to seat. Sometimes there was a bolster at either end as well which was made from a single length of material simply gathered at the ends. Daybeds were either given a single bolster or a set of cushions of graduated size. Because hardly any eighteenth-century sofas still have their original cushions, we tend to think that they weren't meant to have any and were therefore rather formal and uncomfortable. This would not have been true.

Around 1750 a new piece of furniture appeared — the window stool — which largely replaced the conventional stool as a form of seating. These were made in pairs or longer sets according to the number of windows in a room. The long, narrow seat usually continued up and over the slightly scrolled ends without a break, although sometimes the ends were upholstered separately.

Despite the fact that Chippendale stressed that the upholstery covering should be the same as the curtains and wall-hangings, there is plenty of evidence to show that, apart from a co-ordinated color scheme, this was no longer always the case, even in state rooms. Damask upholstery was happily combined with curtains of worsted material and a brightly colored carpet. For everyday use, protective case covers might have introduced another pattern into the room. By 1750 walls were frequently hung with wallpaper instead of fabric. Early papers were intended to look like material, but before long more varied patterns were available. Imported Chinese wallpapers with patterns of exotic flowers and birds were particularly popular, even when the general theme was not Oriental. Chinoiserie patterns were produced by English makers.

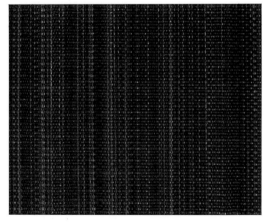

*A*lthough by this period most side chairs had open-carved backs, upholstered backs were not unfashionable. The tapestry covers (left) indicate that this chair was a parlor rather than an eating chair. Tapestry was too expensive to risk soiling with food and drink.

A new material, woven horsehair, was considered very fashionable during the second half of the eighteenth century, particularly for dining chairs. Available today in a wide range of colors and patterns (above), it is rather expensive, but worth the investment.

COVERING A BOX CUSHION

If you are re-covering an existing seat, you can use the old cover as a pattern. Otherwise take some careful measurements — including 12mm (¹/₂in) seam allowance all around. As the cushion will only be visible from the front, a zipper at the back can extend around the sides to allow room for the cushion pad to be inserted. The front and side border can be cut in one piece, and the back border, into which the zipper is inserted, should be made up of two pieces with a 12mm (¹/₂in) seam allowance all around each piece.

To insert the zipper, sew the back border sections together along the seam, using basting stitches only for the zipper placement line. Press the seam open and pin the zipper over the basted seam. Tack the zipper in position and then machine stitch in place. Pull out the basting stitches.

Sew the border pieces together to make a continuous circle. With right sides together, pin the border to the top of the cushion, easing at the corners. You may need to snip the corners to ease the tension. Sew in place and repeat for the cushion bottom. Turn right side out through the zipper opening.

AMERICAN CHIPPENDALE

By the middle of the eighteenth century the most fashionable American furniture was made in Philadelphia and it was there that the "New French" style, was first adopted in the 1760s. On seat furniture this basically rococo style was characterized by cross-bow or serpentine cresting on chair backs, elaborately carved and pierced splats, more elaborate carving on the seat rails and legs and the almost universal use of the claw-and-ball foot. Initially a lot of chairs were made in a transitional style where only one feature was used. For instance, an otherwise perfect example of a Queen Anne chair was given a cross-bow cresting or perhaps a delicate riband splat. Because Thomas Chippendale's pattern books were a great source of inspiration for makers in the new style, it is usually referred to today as "American Chippendale."

The majority of chairs made in Chippendale style had wooden backs, some had drop-in seats, but many had over-the-rail upholstery. The larger lolling chairs with upholstered seats and backs and padded armrests which were so fashionable in England were much less common in America, although examples with Marlboro legs were produced by one of the many European craftsmen, Thomas Affleck, who had emigrated from London to Philadelphia in 1763.

The popularity of comfortable easy chairs was sustained and sofas gradually became quite common. The shape of upholstery on all pieces was identical to that found on equivalent English furniture. Worsted materials, and to a lesser extent needlework, were still used for easy chairs and drop-in seats, but leather and horsehair were popularly used for over-the-rail upholstery. Single or double rows of closely-spaced brass nails were a popular edging, usually placed over a strip of braid or ribbon the same color as the fabric or, occasionally, the heading of a fringe. Fringes on this type of upholstery were not seen in Europe at this date. The edges of easy chairs and sofas were often welted or piped and sofas sometimes had brass nails.

As in Europe most upholstered furniture was supplied with washable protective slip covers. These were usually of furniture check, dimity (an undyed cotton cloth woven with repetitive

Design Detail

Brass nailing was a central decorative feature on Chippendale-style furniture. It was sometimes combined with piping or fringes.

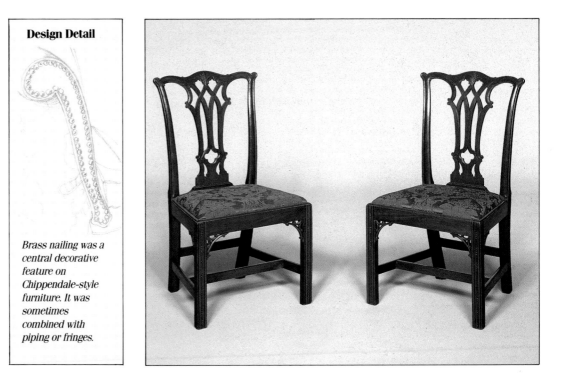

When available, silk damask was the preferred covering for smart chairs such as this one by the Philadelphia maker Samuel Walton (above). As the Chippendale style lasted well into the 1780s, it is difficult to date accurately.

These two chairs (left) are examples dating from the 1760s and 1770s with Chippendale features applied to otherwise Queen Anne-style chairs. Drop-in seats, which were cheaper and easier to upholster, remained quite popular. The chair on the left shows how smart seats looked when covered in damask.

patterns of checks, diamonds, stripes, herring-bone, flowers and so forth), or printed cotton. In ladies' bedrooms and parlors the slip covers often matched the curtains or bed-hangings. Whereas in Europe case covers were quite plain and close-fitting, in America they had a deep ruffled skirt and were sometimes bordered with fringe, another feature not seen in Europe and which indicates that they were meant to be decorative as well as functional.

The Chippendale style remained popular in America until the 1780s, much later than in England. This was due largely to the War of Independence, which effectively blocked the transport of European fashions across the Atlantic for over a decade.

*A*rchitecturally American houses were less ornate than those of equivalent wealth in Europe and the solid respectability of Chippendale furniture suited them perfectly. Easy chairs (above) were almost invariably found in American parlors.

*T*his eighteenth-century furnishing fabric from Williamsburg (left) has a surprisingly modern look. Its lack of sophistication has an appeal of its own.

NEO-CLASSICISM

FRENCH NEO-CLASSICISM
·
NEO-CLASSICAL DESIGNERS
·
FEDERAL FURNITURE

Hepplewhite furniture in a
Federal interior

FRENCH NEO-CLASSICISM

The decorative arts in Europe during the last 40 years of the eighteenth century were dominated by neo-classicism, a style based on the antique architectural ornament of the ancient civilizations of the Greeks, Romans, Etruscans and Egyptians. Antique furniture was not actually copied — not at first, anyway — but existing forms were decorated with classical motifs such as urns, lions' masks, rams' heads, laurel wreaths, classical columns, Vitruvian scrolls, anthemions (honeysuckle flowers), paterae, husks and Greek key patterns.

· REACTION AGAINST ROCOCO ·

In France particularly the neo-classical movement is seen as a reaction against the rather frivolous rococo style, although at first classical motifs were happily combined with basically rococo forms. On seat furniture it was only the carved details which changed, the curvaceous shape of the frame remaining the same. Furniture of this type, made between 1760 and 1770, is called "transitional."

By the time Louis XVI came to the throne in 1774 the neo-classical style was fully developed. Although the King himself was really only interested in furniture from a technical point of view, Marie Antoinette, and consequently most of Parisian society, indulged a passion for luxury and ostentation by commissioning extravagant furnishings of immense richness from the highly skilled craftsmen who were drawn to Paris from all over Europe. Until the Revolution in 1789 Paris remained a symbol of high fashion.

Until about 1785 seat furniture continued to have upholstered backs, which were usually square or rectangular, but sometimes shield- or medallion-shaped. The front seat rail was often rounded and the legs were straight and tapering, carved with either straight or spiral flutings. The frame was usually carved all over, often with running patterns, and the seat rail was quite deep. A universal feature was a knee block at the junction of the legs and the seat rail carved with a small rosette. The arms swept down from the back to join

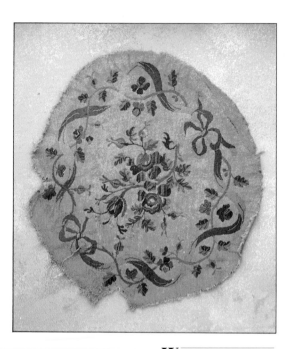

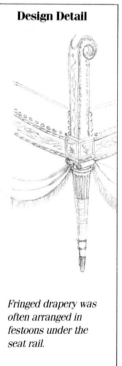

Design Detail

Fringed drapery was often arranged in festoons under the seat rail.

While luxurious silks were the most fashionable upholstery fabrics, the middle classes, particularly in the provinces, continued to use needlework seat covers (above).

Designers took great pains to match the covers as well as the framework of the chairs and sofas with the rest of the room (left).

Striped materials look particularly attractive with the straight and carefully balanced contours of Louis XVI chairs. It is interesting to see how well these light and pretty chairs (right) fit into this rather ponderous interior.

NEO-CLASSICISM

THE IMPORTANCE OF ROME

The greatest source of inspiration for neo-classicism was Rome, where artists, architects, historians and archaeologists from all parts of Europe gathered to study classical remains. Although Italy provided so many of the decorative motifs used by neo-classical designers, Italian furniture makers themselves were surprisingly slow to adopt the new style. Throughout Europe there was a general tendency to use French designs as models for state furniture and English patterns for domestic pieces.

their supports, at first about two-thirds of the way along the seat, but later level with the front. The majority of chairs were either gilded or painted white or pale pastel colors.

Upholstery became more luxurious. The padding of the back was no longer concave, but was thickly and evenly distributed. The seat was a little thicker too and its edge was clearly defined by a line of piping or gimp. Although brass nailing was still used, silk trimmings came back into fashion and lengths of braid were sometimes set an inch or so in from the edge of both the seat and back to give a panelled effect. Festoons of fringed drapery running underneath the seat rail were a new feature and were evidently very fashionable. However, few examples have survived.

· TEXTILES AND DECORATION ·

The most fashionable material for upholstery was silk. The most luxurious covers were embroidered,

*T*he design of this Lyons silk dating from 1770 (above) is closer to the rustic patterns seen on toiles du Juoy *than to contemporary silk fabric.*

*M*ore money was usually spent on beds than on seat furniture. The swags and tassels of this canopied lit à polonaise (right) are typical of the sort of French drapery which Hepplewhite and Sheraton attempted to promote in England.

*T*he Louis XVI style has been consistently popular for reproduction furniture such as the high-quality pieces seen here (left) upholstered in a typical striped and floral pale silk. The lit de repos is properly supplied with both bolsters and cushions. The lyre-backed chair to the right, with its circular seat, is a form of seating popular after 1785.

The upholstery of this almost Directoire-style chair (left) shows the beginnings of the Empire fashion for sharp-edged seats and armrests and box-shaped backs.

This marvellously rich Louis XVI silk damask (below), with its wheat ears and classical garlands, has been perfectly reproduced. Increasing numbers of textile companies are specializing in the exact reproduction of old silks.

mostly with floral patterns, but damasks, moiré, painted and printed satin, lustrings and other striped and patterned silks were also used. White and pale colors were popular, particularly for summer furnishings. Velvet was used in winter.

It was still fashionable to co-ordinate the designs of chairs and sofas with elaborate wall panelling and window curtains. Tables and carcase furniture too were highly decorative, and ormolu mounts, parquetry, lacquer and porcelain plaques were incorporated in the design.

The types of chairs and sofas popular under Louis XV were joined by a new desk chair, the *chaise en gondole*, on which the arms were continuous with the rounded back.

· THE CRAFTSMEN ·

In the provinces a great deal of furniture was still made in Louis XV style, although important châteaux were furnished by Parisian craftsmen. Virtually all makers had access to a growing number of pattern books such as Diderot's *Encyclopédie*, published in parts between 1751 and 1772. It gave detailed instructions for both design and methods of construction. This type of book was extremely influential in spreading the newest fashions.

The most famous *menuisiers* of the period were Jean-Baptiste Sené and Georges Jacob. Jacob was the first important maker in France to use solid mahogany for seat furniture. Around 1785 he began making chairs with pierced wooden backs to imitate lyres, wheatsheaves and woven basketwork. The back legs of many of his chairs were of saber form. Although he preferred to use mahogany, sometimes these chairs were made in beech and then painted or varnished. The upholstery of the seats followed the same lines as other neoclassical chairs. Generally these chairs are more closely associated with the Directoire period (1795-99), which immediately followed the Revolution.

NEO-CLASSICAL DESIGNERS

*H*aving spent four years studying classical art and architecture in Italy during the 1750s the architect Robert Adam introduced Neo-classicism to Britain. On his return to London he joined his family's already established practice and soon acquired a number of wealthy and fashionable clients. A great believer in unity of design, he was responsible for all aspects of interior decoration from the ceiling design to the carpets and in many cases for the exterior architecture as well.

After an early transitional phase, when he applied classical motifs to existing Palladian and rococo forms of furniture, Adam designed chairs and sofas which in most respects resembled Louis XVI models, but which were generally lighter and more delicate. Oval-, square- or shield-shaped backs, round or straight-edged seats and fluted, tapering legs of square or circular section were all standard features. The backs were either upholstered within a carved frame or had a decorative wooden splat. Many pieces were carved and gilded, but on some the decoration was painted on. Alternatively, designs were inlaid in woods of contrasting colors on a light-colored background such as satinwood. His favorite motifs were anthemions, palm leaves, bellflowers, grotesques and continuous scrolling foliage, which he called

*B*rocade (above) was used to furnish only the very grandest State rooms. White and various shades of gold were the most popular colors.

*T*his delicate Robert Adam satinwood chair at Osterley Park (right) is perfectly complemented by its white-painted, Chinese taffeta squab cushion.

52

Design Detail

Patterns on the silk cushions at Osterley repeated those used by Adam on the carpets and other surfaces. The cushions were very fat.

*T*he bolsters on this elegant sofa designed by Adam in 1762 (left) look too small to our eyes in proportion to the rest of the piece, but still make all the difference to its over-all appearance. The seat has very closely spaced shallow-buttoning, although the sides and back are quite smooth.

"flowing rinceau." The upholstery was slightly thinner than that seen on French chairs and the edges were less well defined. Sometimes they were slightly rounded or even flattened off. His chairs with wooden rather than upholstered backs often had caned seats fitted with a bordered squab cushion piped on both edges.

Adam worked in conjunction with many leading cabinet makers and upholsterers of the day such as Chippendale and John Linnell; they too made some fine furniture in the neo-classical style. However, the bulk of furniture made during this period was much simpler. There was already a good tradition in England for solid domestic furniture and those makers who had previously found Chippendale's *Director* so useful, now turned to other pattern books, most notably George Hepplewhite's *The Cabinet-Maker and Upholsterer's Guide*, which was published posthumously by Hepplewhite's wife in 1788.

· GEORGE HEPPLEWHITE ·

One of the *Guide*'s greatest advantages was that it included designs for all types of domestic furniture from fashionable French sofas to simple washstands. Its success was so great that the name Hepplewhite is now commonly given to all English mahogany furniture made during the 1770s and 1780s, despite the astonishing fact that no single piece of furniture made in his workshop has ever been identified.

The best-known designs in Hepplewhite's *Guide* are those for parlor or dining chairs, with or without arms. Most had straight, rather thin legs, either square or round in section, which tapered from the knee block down to feet ending in a small plinth. A few turned slightly outwards towards the foot. Most of the deli-

53

The richness of some of Adam's very grand interiors can be seen here in the Red Drawing Room at Syon House, near London, where the walls are still hung with their original brocaded Spitalfields silk. The magnificent set of gilded chairs, sofas and stools stand against the walls.

cately carved and pierced backs were shield-shaped, but some were square, oval or heart-shaped. Later editions of the *Guide* showed a greater number of square or rectangular backs. A few backs were upholstered, but most had splats carved with classical urns, drapery, wheat ears, vases and so forth. The Prince of Wales feathers were a popular late feature.

The seats, which were either straight, bowed or serpentine at the front, had single or double rows of nails. A few of Hepplewhite's designs showed swags or festoons of nails. Unlike Adam chairs, where the seat rail was carved, Hepplewhite chairs were covered over the rail. Where an upholstered back occurred, Hepplewhite indicated slightly concave padding. Padded armrests were virtually flat.

Hepplewhite also designed a number of stools, including some with gently curving cabriole legs. The upholstery was similar to that shown on the chairs, although one had a short fringe under the seat rail. A selection of window stools with similar upholstery was also shown. One was shallow-buttoned, while a couple showed small festoons of fabric draped underneath the front rail. One set of drapes was fringed and both had silk tassels hanging between the swags.

Hepplewhite sofas and settees were particularly delicate and sometimes had baluster-shaped legs. Various French types of sofa were made too, in particular the *confident*, which was made in one piece, but in three sections.

A great deal of Hepplewhite's seat furniture was constructed in mahogany, but some was fashionably painted to co-ordinate with the room, or japanned, or made in satinwood with painted or inlaid decoration.

A surprising feature of Hepplewhite's *Guide* was the appearance of rococo designs. It is mostly apparent in rather elongated cabriole legs, but armchairs of totally Louis XV type, although thinner and more delicate, were also made. Today these armchairs are called "French Hepplewhite."

· THOMAS SHERATON ·

Another designer whose work was particularly influential during the 1790s and early in the nineteenth century was Thomas Sheraton. Most of his designs were similar to Hepplewhite's, but those for upholstery were far more lavish, particularly drawing-room furniture. Sheraton's designs for chairs and sofas show a greater angularity; straight seats and square or rectangular backs, whether wooden or upholstered, appear more frequently. For dining chairs he recommended mahogany, but drawing-room chairs he believed should be painted

54

Many neo-classical chairs, such as these fine mahogany Hepplewhite-style examples (left), had cane seats. The modern squabs, with their neatly piped borders and shallow-buttoning, carefully follow the line of the lower back rail.

Such lavish upholstery as shown on this Sheraton design for a chaise longue (right) can seldom have been executed and certainly could not have stood up to much use, particularly when upholstered in fine silk.

A Hepplewhite design for a classic window stool (left) has a silk-tufted cover with muted stripe. The fabric looks as if it was intended to represent horsehair.

These generally lighter neo-classical armchairs have correspondingly flatter upholstery than their mid-century prototypes.

Painted chair frames (below) often had upholstery of matching color.

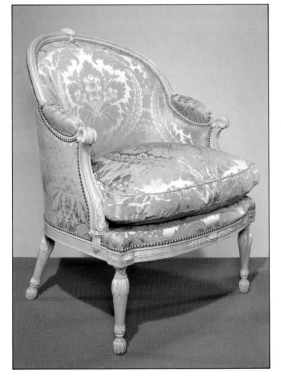

CLOSE-NAILING

This is a difficult but attractive way of finishing a border. It is generally done by eye and is a skilled procedure — just one nail out of place can ruin the final effect. Dome-headed nails. usually with a 12mm (½in) head and 12mm (½in) shank, are tapped in so that each head just touches the one next to it. It is important to place the tacks at the right intervals to allow the shanks of the nails to fit in between them.

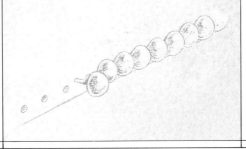

or japanned. Many of the latter were made in beech and had caned seats and squab cushions.

On his designs for fully upholstered pieces, Sheraton indicated a much sharper, right-angled edge on both seats and backs, but even so it was not defined by piping or braid as other pieces were. Some of the designs show loose cushions and bolsters, which by now had a separate circle of material at either end. He showed elaborate drapes, usually fringed, which hung from under the seat rail or were attached to the edge of the seat upholstery. One of his designs for *chaises longues* – a piece of furniture which was to become very popular during the Regency period and which he described as something "to loll upon after dinner" – has additional drapes on the inside and outside of the back and on the outer sides of the arms.

As in France the grandest neo-classical furniture was upholstered in fine white or pale-colored silk, which could be either plain, figured, painted, printed or embroidered. Sometimes medallions or borders of printed silk were bought ready-made and sewn by the upholster-er on to a plain background. Damask was still popular, but the patterns were smaller than before. Brass nailing, piping, colored braids and short tufted fringes were used for decorative edging. Horsehair and leather continued to be popular for dining rooms; blue and red were the most fashionable colors for fabrics. Hepple-white recommended that leather upholstery should be tufted.

On all materials large patterns were re-placed by smaller repetitive designs; ribbons, sprigs of flowers, urns and other classical devices were popular motifs. Plain stripes and patterns composed of alternating stripes and flowers were also common, particularly after 1780. The same sort of patterns were seen on the printed cotton loose covers that became fashionable for bedrooms and parlors. Some-times the covers were close-fitting and the up-holstery underneath was simply covered with linen or calico. These were obviously intended to be permanent. Alternatively they were quite baggy and sometimes reached right to the floor. Presumably these were really a decorative form of protective case cover. Both types were usually made of the same material as the cur-tains and it was still common to have two identical sets. One could replace the other while the covers were being washed. Usually the seat edges were piped and the border edged with braid. By the end of the century cotton printers were producing pre-shaped seat covers with borders to match.

55

FEDERAL FURNITURE

By the end of the eighteenth century, American furniture was still very much dominated by the designs of Chippendale; the influence of Adam and the neo-classical French style had been late in coming because of the hostilities resulting from the Revolution. However, by 1790, these fashionable designs were taken up enthusiastically, spurred on by the French association; the French had aided the breakaway nation in its fight for independence.

The influence of the first capital at Philadelphia was eventually taken over by Boston and then New York, where furniture was exported abroad in return for the desirable cargoes of the southern continents. Baltimore was another center of French-influenced Federal furniture.

Much of the furniture produced during the Federal period was precisely copied from published pattern books such as those of Hepplewhite and Sheraton. Their unpretentious designs were appropriate for the republican ideals and lifestyle. Some of the patterns were reprinted in widely distributed publications such as the *Philadelphia Journeyman Cabinet and Chairmakers' Book of Prices*. At the same time the influence of Robert Adam was seen in the use of colorful woods and decorative inlays and a partiality for painted decoration and caned seats for chairs and settees.

Because of this English influence, many Federal-period chairs, settees and sofas cannot be easily distinguished from those made in England. There was the same preference for delicately carved and pierced chair and settee backs and narrow, tapering legs. Seats, on fully upholstered pieces, were gently curved or serpentine at the front and the sides curved inwards towards the back. Over-stuffed chairs had rather flat, even upholstery with slightly rounded edges. Double rows of closely spaced brass nails were a common feature and some makers directly copied Hepplewhite's decorative designs for little swags and festoons of nails. Occasionally the nails were tacked into the heading of a long fringe, and on winged chairs and sofas the arms were piped. Neither feature really caught on in England.

Prosperity soon followed the resolution of

THE MARTHA WASHINGTON CHAIR

In New Hampshire, a chair unique to America was produced between 1800 and 1810. This was the Martha Washington, named after the President's wife. It had an unusually tall, straight, upholstered back and wooden arms which continued down into the front legs. Large numbers are still in existence.

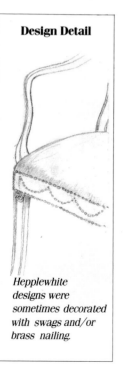

Design Detail

Hepplewhite designs were sometimes decorated with swags and/or brass nailing.

Although its arrival was delayed by the American Revolution, the elegant sophistication of neo-classical furniture (left) was immediately popular in America. Lighter colors introduced a touch of femininity to rooms.

The use of small-patterned printed cottons, as seen on this Martha Washington chair (above), became increasingly common, particularly for bedrooms. The brass nails are almost too smart for the fabric.

56

57

the conflict, and New York emerged as the leader in furniture design and production. Duncan Phyfe established a workshop there and catered to the fashionable wealthy people both in New York and throughout the states, particularly in the south. Satinwood was popular at the time, but Phyfe worked in mahogany, the darker wood giving authenticity to his neo-classical styles.

Though the use of English pattern books was universal, each furniture-making center maintaining its individuality by favoring a particular design or decorative technique. A variety of woods was used. Birch, maple, cherry, walnut and American plane were alternatives to the popular mahogany. Some details were slightly changed. For example, Thomas Sheraton's designs were the basis of the popular New York design for an occasional chair. Small, painted, with cane seats, they were made in sets with a matching settee and thin, rather flat squab cushions, often bordered with twisted cord and matching tassels at the corners.

· TEXTILES ·

Satin, velvet and silk damask were the first choice of fabrics in the houses of the very wealthy. Crimson and blue seem to have been the most popular colors, followed by green and then yellow. The less well-off were content to have their furniture covered in patterned worsted fabrics such as moreen and harateen, usually in red, blue and green. Leather was replaced by black, red and green woven horse-hair, particularly for dining chairs.

For slip covers and gathered skirts, cotton chintz, sometimes glazed, with either woven or printed patterns, was popular. Multi-colored small floral patterns, mostly with a strong vertical emphasis, replaced the large monochrome designs fashionable during the previous period.

Dimity, used in Britain for clothing, was brought into bedrooms as bed hangings and curtains. Blue and white checked linen was made into case covers. In retrospect, this use of simple basic fabrics and patterns shows an originality that is typical of the pioneering spirit of early American society.

THE EMPIRE AND REGENCY PERIODS

FRENCH EMPIRE STYLE

·

REGENCY FURNITURE

·

BIEDERMEIER STYLE

·

EMPIRE STYLE IN AMERICA

*David's famous portrait of
Madame Récamier*

FRENCH EMPIRE STYLE

espite the upheaval of the French Revolution and the Terror that followed it, there was a continuity in furniture design; the archaeological approach to classical design had actually begun before the Revolution, but it did not come to fruition until the nineteenth century. The neo-classical themes had appealed to the revolutionaries; this was the time of the famous painting of Madame Récamier by David showing her lying on a daybed. Even her dress and hairstyle shows Grecian influence. It wasn't until Napoleon Bonaparte became Emperor in 1804 that France's skilled craftsmen could once more practice their arts. So many of their patrons had been swept away with the Revolution.

Napoleon's efforts to distance himself from the Bourbon kings and identify his reign with the splendor of the Roman Empire meant that he commissioned whole new interiors for the royal palaces. The most important designers of the Empire period were Charles Percier and Pierre Fontaine, whose furniture was made by Georges Jacob and later by his sons Georges and François. After Georges' death, François traded under the name Jacob-Desmalter.

The rather theatrical background to the Empire style that Percier and Fontaine were commissioned to produce for Napoleon and Josephine was recorded in their 1801 publication *Recueil des décorations intérieures*. Their designs for seat furniture were based closely on Greek models and included both single- and double- scroll-ended couches, X-frame curules and Greek *klismos* chairs. All three became universally popular throughout Europe and in England. Italy particularly, ruled by Napoleon's relations, took to the style wholeheartedly.

Popular decorative motifs included laurel wreaths, stars, fasces (bundles) of arms, dolphins, swans, animals' heads and feet (particularly the lion's paw), eagles' wings, lyres and various other foliate designs. During Napoleon's campaign on the Nile, Egyptian motifs, particularly the sphinx, became popular.

Some of the grandest furniture was made in carved and gilded wood, but figured veneers decorated with rather sparsely distributed ormolu mounts was more common. Marquetry virtually disappeared. Because of the continen-

The rich silk velvet upholstery of Napoleon's chair (left) matches the table cover and the decoration of the chair frame.

The "gold" color woven into this reproduction Empire fabric makes it sufficiently grand for these carved and gilded armchairs (above).

Design Detail

The corners of seats were upholstered in a sharp, well-defined edge.

Empire chairs are peculiarly difficult to upholster in anything but the correct fabric. Empire designs (left) were already being reproduced by 1850.

Fringes of clearly defined shape, such as this (below), were fashionably placed underneath the front seat rail. Very few have survived still in place.

tal blockade by the English navy, mahogany from Central America, which had been used a great deal during the post-revolutionary period, was in short supply and was often replaced by light-colored indigenous *bois clairs* such as elm, ash, oak, walnut, yew and fruitwood.

A watered-down version of Percier and Fontaine's Empire style appeared in La Mesangere's *Collections des Meubles et Objets de Gout*, which was published in magazine form from 1802 until 1835. These influential designs were widely used by makers supplying furniture to the bourgeoisie.

· CHANGES IN SHAPE ·

Some early Empire chairs and sofas retained the straight legs with spiral flutings seen on Louis XVI examples, but these were soon replaced mostly be the simple concave saber legs seen on *klismos* chairs, but occasionally by animal legs. The back legs of nearly all seat furniture were also of saber form and often formed a continuous curve with the uprights of the back. The back either curved over backwards at the top in a scroll or was composed of a concave horizontal rail extending outwards beyond the uprights. Wooden arms were usually an extension of the front legs. Sofas had similar but slightly shorter legs, while couches had a broad seat rail and large outward-curving feet.

Seat upholstery was severe, often box like, in form, with rigid straight edges, flat surfaces and sharp, right-angled corners. Even the back was sometimes of box form, although occasionally it had smoothly rounded upholstery in sharp contrast to the seat. On *chaise en gondole*, which became particularly popular at this time and were essentially rounded in shape, this was a necessity.

· DIVERSITY OF FABRICS ·

Upholstery fabrics became much simpler. Silk remained the favorite for grand furniture and was woven or printed with formal classical patterns such as stars, laurel wreaths and medallions. White, yellow, gold, pale green and deep pink were fashionable colors. Broad decorative braids, often gold or of a contrasting color to the material, emphasized the straight lines and box shapes of the stuffing. Short, plain fringes sometimes hung around the underneath of the seat rail. Scroll-end couches were invariably fitted with large bolsters, the ends decoratively pleated into a central rosette, sometimes with a tassel attached. Sofas had loose cushions leaning against the arms, but no cushions along the back. Less expensive furniture was upholstered in small patterned *toiles* (printed cottons), cotton damasks, plain velvet or worsted fabrics. Decoration was again supplied by braids and fringes.

As official designers for the Emperor, Percier and Fontaine were often called on to create temporary settings for him while he was on his travels and for this they often used striped draperies arranged in tent fashion. Consequently beds, and often whole rooms, were permanently hung in this way and striped material became a popular covering for upholstery. Striped wall hangings and papers and elaborately draped curtains were also a feature of the period.

The Empire style lingered on in France long after Napoleon's exile in 1815, but was generally heavier and plainer than before. New political circumstances and the rise of the middle class under the Bourbon Restoration and the July Monarchy, meant that there was much less demand for very opulent furnishings. Late Empire pieces were characterized by the use of highly figured woods sparsely inlaid with brass, pewter and ebony. Upholstery followed the same lines, but the choice of materials and colors was plainer and more somber.

Despite the Revolution, grand interiors were still being commissioned under Napoleon. The splendid white and gold satin upholstery with its square edges and gold braid (left) epitomize the Empire style.

Classical motifs appeared on even the simplest chairs. On this otherwise plain piece (below) rather flat ormolu mounts are combined with carved mahogany rams' heads.

61

REGENCY FURNITURE

he trend for simplified versions of grand furniture in the French taste, which the pattern books of Chippendale and Hepplewhite had helped to promote during the eighteenth century, continued during the Regency period. Although George IV was only Regent from 1811 to 1820, the term Regency is used to refer to the years lasting from approximately 1800 to about 1830.

In line with France, neo-classicism in Britain also entered a second, more archaeological, phase, although even the most sophisticated furnishings did not match the ostentatious grandeur of their French Empire equivalents. Despite the extravagant interiors created for the Prince Regent in the Brighton Pavilion and elsewhere, the general trend among the aristocracy was for small, intimate villas in "picturesque" style. The simple domestic furniture that these smaller houses demanded inevitably closed the gap between furniture made for the wealthy and that which was affordable by the middle classes. Smaller houses and a more intimate way of life meant greater informality and furniture began to be placed away from the walls more towards the middle of the room.

In England the furniture closest to the Percier and Fontaine Empire style was that made for a wealthy amateur architect, Thomas Hope, whose designs were published in 1807 in a book titled *Household Furniture and Interior Decoration.* Although Hope's designs became well known, they were too archaeologically correct for popular taste and other publications that showed a watered-down, practical version of the Empire style were probably more influential on general furniture production. Included in these were Thomas Sheraton's *Drawing Book, 1802,* Ackermann's *Repository of Arts* (published in parts from 1809 to 1828) and George Smith's *A Collection of Designs for Household Furniture and Interior Decoration,* 1808. Smith's book was particularly useful because it showed the full range of furniture required for all the main rooms of a house.

The same types of seat furniture that were popular on the Continent were fashionable in England — *klismos* and scroll-back chairs, both with saber legs, curules and scroll-end sofas and *chaises longues.* They showed the same rectilinear, architectural outlines and sparse use of classical ornament such as anthemions, lion's paw feet, Greek key patterns and so on. Marine emblems, such as dolphins and twisted rope, were popularized after Nelson's victories and the use of Egyptian motifs increased after the Battle of the Nile. Rush-seated fancy chairs, carved and painted to look like bamboo, became fashionable following completion of the chinoiserie interiors of the Prince Regent's Brighton Pavilion.

· DECORATION AND DETAIL ·

Cane seats were quite popular on chairs of all types. *Bergère* armchairs — quite different from French *bergères* — which were caned on the back and arms as well as on the seat, were frequently seen in libraries. Stools, particularly those of curule form, were a familiar sight in all rooms and were often placed in window recesses. The close relationship between a building and its surrounding landscape, which was such an important feature of picturesque villas, gave window seats added importance and paintings show that they were sometimes built into the window panelling and fitted with squab cushions.

Although some Regency furniture was gilded, the majority was dark in color. Plain maho-

Design Detail

The circular ends of bolster cushions were sometimes decorated with a rosette.

The imposing interiors (left) of the Prince Regent's London home, Carlton House, equalled those of Napoleon in their grandeur. Note the large and rather squashy cushion on the otherwise severe sofas.

Block-printed cottons with a combination of floral and leaf patterns became increasingly popular for bedrooms, although usually for loose covers not fixed upholstery, as above might indicate.

Although the striped silk upholstery of these pieces by Henry Holland (left) is not original, it is suitable for their classical shape.

Caned bergère chairs with leather upholstery (above) were a standard feature of Regency libraries.

gany and rosewood, and ebonized, japanned or painted beech were all common and were often decorated with inlay of brass or with incised and gilded patterns. On grander pieces, small ormolu mounts were applied and reeding was a popular form of carved decoration.

Dark frames required strongly colored upholstery and deep pink and crimson, blue, green and deep yellow were all popular. Nearly all designers suggested using plain materials. In smart furnishing schemes the seat upholstery reflected the elaborate fringed and tasselled curtains and valances, which framed the windows. The stuffing followed French lines; high, square-edged seats had borders and panelling of braid in gold or contrasting colors. Long, sometimes floor-length, fringes hung from the seat rail and were now seen on the back as well as the front and sides, an indi-

cation that the chair was intended to be seen from all angles, not just against the wall.

The design books show that both couches and chairs often had drapes under the seat rail and on the back and arms, although their appearance was much looser than the drapes seen on the eighteenth-century pieces. In Grecian style completely loose drapes or mantles were casually hung over the backs and arms.

Bolsters were an essential feature of sofas and couches and often had additional, rather flat, rectangular cushions that leant across them against the arms. The ends of bolsters were made from a separate piece of pleated material, which radiated out from a central rosette or tassel.

All cushions, including squabs, were made from two flat pieces of material without a bor-

FASHION IN DRESS

This had its effects on seat furniture. The simply cut, narrow dresses à la grecque or "Empire" gowns which were introduced during the 1790s led to narrower chair seats and made reclining a much easier activity for genteel ladies. Couches and chaises longues, which could of course be used for sitting as well as reclining, therefore became more common and were fashionably placed on either side of the fireplace or in the center of the room next to a pedestal or sofa table.

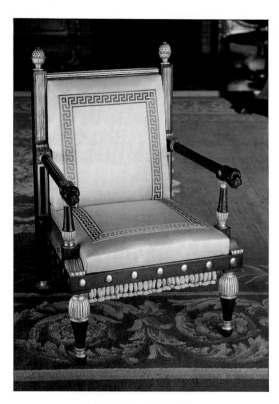

The Regency style reached all centers of furniture production, even those in the colonies where these painted chairs (right) were probably made. Silk rather than velvet squabs would lighten their slightly heavy proportions.

The tasselled fringe and the broad decorative braid with its bold key pattern have been accurately reproduced on this famous Thomas Hope chair (left). The strong color is typical of the period.

The ebonized background to the classical decoration of this klismos armchair (above) was intended to reflect the color and decoration of antique Greek pottery.

The notoriously bizarre and extravagant chinoiserie interiors of the Prince of Wales' Brighton Pavilion (right) were mostly furnished with magnificent Empire furniture.

der. Twisted cord was a popular edging and was sometimes looped or knotted at the corners as a substitute for tassels. Squab cushions had a loose effect and appeared to droop slightly over the corners of the seat. A new feature was the open-ended cushion in which the ends were loosely tied at intervals with cord or buttons allowing a contrasting lining to show through the openings.

Silk, satin, cotton velvet and plain woolens were popular materials and leather was still used for libraries and dining rooms. George Smith mentioned that decorative borders were often printed onto leather chair seats. It is interesting to note that plain striped fabrics, which for years have been used for the re-upholstery of Regency furniture, was not used in England for fixed upholstery until the 1820s, although striped cotton had been popular for case covers since the 1780s.

Wood-block printed chintzes with large multicolored flower and leaf patterns, often combined with architectural ruins and columns, were quite common in bedrooms from early on in the period. After 1820 furnishing cottons were mostly roller-printed and it was then that striped patterns, often interspersed with bands of flowers, began to be more popular. Illustrations show a wider use of decorative loose covers in drawing rooms as well as bedrooms, a practice that was to be a characteristic feature of domestic furnishings until the end of the century.

From about 1825 furniture began to lose its former elegance. Dark colors were replaced by lighter woods such as maple and satinwood, and large and rather coarsely carved classical motifs replaced inlay. Chair legs were mostly straight and heavily turned or fluted and the back yokes were broader and fashionably carved with scrolls and volutes. Sofas had short, stumpy, turned legs and a large cresting, often of honeysuckle form, in the middle of the back. Popular pattern books, such as Nicholson's *The Practical Cabinet Maker, Upholsterer and Complete Decorator* (1826) and George Smith's *Cabinet Maker and Upholsterer's Guide* of the same year, show the beginnings of this coarsening of design and the movement towards the eclectic revival styles which were soon to follow.

BOLSTER CUSHIONS

If you are using a patterned fabric to make your bolster, note that the pattern of the bolster should correspond to the arm against which it will lie. If it has stripes, for example, be sure to line up the stripes exactly so that they follow across the arm and continue over the cushion. As well as the upholstery fabric, you will need a zipper to run almost the full length of the cushion, some piping for the ends and enough muslin to make reinforcing circles for each end. Covered buttons are needed to decorate the rosette ends.

Measure up for the cylindrical section and cut one piece leaving an allowance of 12mm (½in) for seams all round. For the gathered ends of the cushion, cut two long strips of material, the length equal to the width of the body and the width half the diameter of the end with an allowance for the seams.

Begin by making the cylindrical piece. Machine stitch at each end and baste stitch the opening where the zipper will be inserted. Press the seam open and pin the zipper over the basted seam. Stitch by hand with a neat small backstitch. Take out the basted seam. Machine stitch and press open the seams on the two strips for the ends. Tack and stitch them to the reinforcing circles of muslin, snipping the edges to ease the tension. Gather the unsewn edge and pull the gathering thread tight so that it lies flat against the muslin circle. Secure the ends of the gathering thread and cover the exposed ends with a covered button or tassel. Attach the piping (see page 110) around the edges of the circles, again snipping to ease tension. Machine stitch the circles to the cylindrical bolster. Turn the bolster cover right side out and insert the cushion pad.

65

BIEDERMEIER STYLE

French Empire furniture was greatly admired by the rest of Europe and its influence was widespread. In Germany and Austria elements of the Empire style were evident in the distinctive domestic furniture made in what is now called the Biedermeier style.

Biedermeier was very much a middle-class movement in which the bourgeoisie demanded elegant yet practical and inexpensive furniture for their modest and unpretentious homes. The word Biedermeier derives from the names of two comical characters of the time. The wars with Napoleon had left the Austrian monied classes relatively impoverished and the simplicity of the Biedermeier furniture suited the more austere times. Indeed, the most obvious feature of Biedermeier furniture was its simplicity. The shape of each piece was basically geometric and although curves were used, they were invariably of simple, concave form. Ornament was restrained and limited to simple inlay or small gilt bronze mounts of classical form, such as sphinx heads or lions' paws. Many pieces were totally undecorated, the grain of the wood being the most important feature. There were some regional differences. In Austria light-colored fruitwoods were popular, but in northern Germany darker woods were preferred, particularly mahogany.

The best-known Biedermeier designers were Josef Danhauser, who ran an immensely influential *Etablissement für alle Gegenstande des Ameublements* in Vienna, and Karl Friedrich Schinkel, who worked in Berlin.

Houses of the period were sparsely furnished and more informally arranged than before. Usefulness and comfort were more important considerations than appearance and furniture was no longer designed to stand rigidly against the wall. The actual types of furniture were quite limited. A deep, straight-fronted sofa with a plain, flat-surfaced frame and straight or saber legs was the most important form of seating. Sometimes sofas were of the scroll-end type, although the scroll of the arms was slight. The shape was usually echoed in the shape of the feet. Occasionally a solid base with scrolled ends replaced legs.

While simple outlines and minimal decoration obviously reflected the demands of a bourgeois clientele, a great deal of Biedermeier furniture is extremely smart (above).

Design Detail

The scroll end of the sofa arm could be slight, as here, or more pronounced. The shape might be repeated in the leg.

Although based on Empire classicism, Biedermeier developed a unique style of its own. This imposing sofa (left), with gilded carving and elaborate feet, is unusually ornate.

Chairs were generally of two types: one had a rounded back that reached forwards on either side of the horseshoe-shaped seat to form the side seat rails, while the other was of the *klismos* type, but with less pronounced curves. The front legs could be either straight or saber, but the back legs were invariably outward curving.

· UPHOLSTERY METHODS ·

Biedermeier makers were keen on complicated upholstery, although the fabrics themselves were quite simple. Plain or striped silk or velvet, leather and horsehair (the latter often black), were the most common materials. The padding on backs and arms was often thin, but the seat was generally quite deeply upholstered with a sharp right-angled edge. Interesting effects were created by drapes of material or by pleating radiating from a central rosette. A limited use was made of fringes and edges were mostly bordered by narrow gimp. Drapes were often accompanied by silk tassels, which were also seen on the corners of the squab cushions made for cane-seated chairs.

The Biedermeier style is said to have lasted until about 1850, but by then the forms had become clumsier. The best Biedermeier furniture was made before 1830.

67

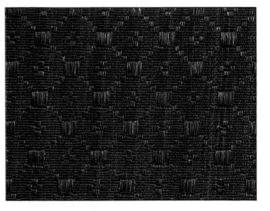

*B*iedermeier makers were admired for their fine upholstery (left). The introduction of sprung upholstery is attributed to the Austrian Georg Junigl. Despite some extravagant designs for draped and pleated coverings, simple striped materials were the most popular choice.

*T*he sharp contrast between light woods and a black ebonized finish was sometimes emphasized by the use of dark patterned horsehair (above).

EMPIRE STYLE IN AMERICA

uring the 1790s large numbers of Parisian craftsmen emigrated to America and took with them their knowledge of the new classical style, which had just begun to develop before the Revolution. Many of them settled in New York, which was fast becoming the most important center for the furnishing industry. Before long they were producing large quantities of mahogany seat furniture in the form of *klismos* chairs, curules and scroll-end sofas and couches, with carved decoration of Greco-Roman motifs. The X-frame curule was popularly adapted to form the base of chairs and settees and scroll-back chairs, with either straight or saber legs, were made in large quantities with cross or lattice backs, or splats of lyre shape. After 1810 brass lions'-paw feet were quite a common feature.

The use of metal mounts for all kinds of furniture was popularized by an influential French maker, Charles-Honoré Lannuier.

Many chairs and settees were made with caned seats and sometimes caned backs and arms as well. Flat squab cushions without borders were edged with twisted cord and had large fancy tassels hanging from each corner. Fully upholstered pieces had the square-edged stuffing seen elsewhere, with wide borders of patterned braid in contrasting colors. Flat fringes or elaborate fringed drapes often hung from the seat rail. Although fatter than squabs, the loose cushions seen on sofas also had no borders, but usually had decorative tassels on each corner.

Silk was the most fashionable form of covering, either plain or with a small pattern. Plain and patterned velvets were also used as well as leather and horsehair. Plain colors with printed patterned borders were used in drawing rooms, but dimity, furniture check and chintzes were still popular for bedrooms.

68

Design Detail

The corners of cushions were decorated with tassels, twisted cord or knots of fabric.

Because so many French craftsmen emigrated to America after the Revolution, furniture in full-blown Empire style (right) was highly fashionable, particularly in New York where so many of them settled. However, in the re-upholstery some distortions do arise, such as on this magnificent chair where the large, soft seat cushion is out of keeping with the severe design.

The grand classicism of French Empire was toned down to suit more modest homes. On this elegant mahogany sofa (left), the transition from neo-classicism to Empire is evident in the combination of Adam-style carved drapery on the back rail and the two curule supports to the seat frame.

THE EMPIRE AND REGENCY PERIODS

NINETEENTH-CENTURY BRITAIN

THE VICTORIANS

·

ARTS AND CRAFTS

·

PROGRESSIVE DESIGNERS

·

EDWARDIAN REPRODUCTIONS

An aesthetic interior by
Kate Hayllar

THE VICTORIANS

ictorian furnishing fashions were to a large extent dictated by the demands of the rapidly expanding and increasingly prosperous middle classes. They wanted to emulate the interest of "society" in past styles, but were unable to afford the real thing. While the wealthy filled their houses with either genuine — and mostly French — antiques, or with high-quality reproductions, the middle classes purchased vast quantities of brand new furniture in the "antique style." The success with which furniture manufacturers interpreted earlier styles varied considerably, and even though historical accuracy was often lacking, most pieces had an attractive and positive style of their own.

The most obvious feature of nineteenth-century interiors was their eclecticism. From the 1830s onwards an enormous number of styles — many of them historical revivals — were popular at the same time, and their simultaneous use within a house, though not necessarily within the same room, was perfectly acceptable. Individual styles were thought suitable for particular rooms: during the early part of Victoria's reign (1837-1901) French styles, particularly rococo, or Louis as it was called at the time, was appropriate for drawing rooms, and Renaissance for dining rooms.

Later in the century that curious curvaceous style that evolved from rococo was joined in drawing rooms by Sheraton, Adam and Hepplewhite. *Painted Adam* and an oddly architectural form of Queen Anne were favored for bedrooms. Gothic was popular for libraries and billiard rooms and "Mooresque" for that newly fashionable and essentially masculine room, the smoking room. Another masculine style was "early English," which was based on English medieval forms. During the 1870s the revival styles were joined by Anglo-Japanese and other forms of spindled, bracketed and usually ebonized "art furniture" as it was known.

For many the essence of early Victorian design was the gothic. This was in part due to the energy of A.W.N. Pugin, who promoted the gothic style in furniture and architecture as a crusade against the frivolity of neo-classical themes. His pattern book, *Gothic Furniture of the Fifteenth Century* (1835), was very influential, although it was taken up mainly in public

buildings, particularly the Houses of Parliament, where Pugin helped with the interiors, and less so in family homes.

· VICTORIAN INTERIORS ·

Co-ordinated furnishings and carefully balanced room arrangement were replaced by a mass of contrasting colors and patterns and a vast concentration of furniture indiscriminately crammed together and over-loaded with ornaments, fans, picture frames, lace doilies and other knickknacks. Green and brown paintwork, dark-colored, large-patterned wallpapers and heavily draped curtains and upholstery combined with poor lighting to give a general impression of gloom. It has been suggested that dark colors were favored because the discoloration caused by air pollution, particularly in urban and industrial areas, was less obvious.

Middle-class ladies were advised on their choice of furnishings by a wealth of "artistic" periodicals. Later in the period they could go to one of the many house furnishers or department stores, where they could view a good

Design Detail

Long fringes were just one of the many trimmings used by the Victorians to decorate their furnishings.

The gothic style, promoted by the designer A.W.N. Pugin as the only true style of architecture, was principally used for large-scale interiors. At Eastnor Castle (above), Pugin revived the folding X-frame chair of State and used the same velvet and gilt-headed nails to cover other pieces in the suite.

A co-ordinated color scheme was used for this 1840 Bristol drawing room (right). The loose cover on the large sofa has cushion covers of the same fabric and also matches the fixed cover on the chair next to it.

range of furnishings of varying price and quality and buy virtually ready-made interiors. The catalogues that some of these firms issued, as well as the trade catalogues supplied to retailers by the manufacturers, have proved very useful to furniture historians, as they clearly show the type of furniture used in the average household rather than the pieces made for individual clients that illustrate the various styles at their best.

As before, manufacturers were greatly influenced by pattern books, although early in the period these were quite scarce and were often re-issued over a considerable number of years. J.C. Loudon's *Encyclopedia of Cottage, Farm and Villa Architecture and Furniture*, which showed a wide range of middle-class furnishings, was first published in 1833 and re-issued virtually unchanged in 1842, 1846 and 1857. It was used by makers in both England and America.

· ADVANCES IN UPHOLSTERY TECHNIQUES ·

Although upholstery was an important feature of interiors, the status of the upholsterer was reduced to that of a mere factory worker in Victorian times.

Greater emphasis on comfort came with the

73

This rococo-style easy chair (above left) has thickly rounded upholstery with pleated fabric between the buttons.

One of the most popular Victorian chairs, the prie dieu (above), was fashionably covered with Berlin woolwork embroidery.

introduction of the coiled spring in the 1830s. To prevent the springs from pushing up through the covering material, much fatter stuffing was placed on top and buttons were often deeply set between them to hold them in place. The thickly rounded effect which this created was repeated on all the edges of the upholstery and soon determined the shape of the framework it covered. It eventually covered the frame itself. Rounded upholstery went well with the curvaceous outlines of rococo furniture and an enormous quantity of seat furniture in this style was made during the following 30 to 40 years.

Alongside the coiled spring, there were other technological advances that came about with the Industrial Revolution, but most of them didn't gain full acceptance in the essentially craft-based furniture trade. Iron, for example, was used for some frames. *Papier mâché* was decoratively used with mother-of-pearl in the frames of chairs and tables. However, the Victorians' love of clutter meant that many innovations were not readily visible.

REPLACING BUTTONS

To replace a button you need a 250mm (10in) needle, some twine and an odd piece of webbing or other thick fabric. Undo the back cover to get at the button hole. Using the existing hole, insert the needle through from the back, thread the button and pass the needle back through the hole. Remove the needle, pull the button into place gently with the two cords and tie a loose slip knot, placing a toggle of folded webbing under it to stop the cord from cutting into the stuffing. Slowly tighten up the knot.

74

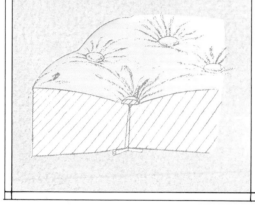

The Victorians were always eager to explore the possibilities of new materials and techniques. Even this brass rocking chair (left), which was shown at the Crystal Palace Great Exhibition in 1851, was covered with buttoned upholstery.

The use of deep-buttoning sometimes interrupted the design of a fabric and all-over patterns proved to be the most successful. The decorative effect of the buttoning here (above) overrides the pattern of the fabric itself.

· NEW TYPES OF SEATING ·

The variety of chairs, sofas and other seats increased considerably during this period. Balloon-back chairs with delicate cabriole legs were the most common form of dining and occasional drawing-room chair until at least 1870. Vesper chairs, sometimes called *prie dieu* or devotional chairs, with short legs and a tall T-shaped back were also popular. A fashionable form of desk chair evolved with a U-shaped seat echoed above by a similarly shaped horizontal upholstered bar. At first this was supported only in the center by a broad strut, but later by a continuous row of turned spindles.

Sofas and couches were made in curving shape and were also often upholstered all over, the only visible wood being on the legs. Sometimes there was a pronounced hump in the middle of the back or a small hump in the center and a large one at each end. *Chaises longues* were made on similar lines, sometimes without the back armrest. From about 1850 onwards the wooden frame was fashionably exposed. Smoothly rounded mahogany frames — described by one furniture expert as having the appearance of being squeezed from a tube — were sparsely decorated at intervals with small carved details.

Sofas and couches were accompanied by a large assortment of fully upholstered lounging chairs, with or without arms. Drawing-room furniture was available in suites, which mostly consisted of a couch, half a dozen single chairs

THE OTTOMAN

A fashionable form of seating from the late 1820s onwards was the ottoman. Originally these consisted of a simple upholstered box with a back and a number of loose cushions. They could have either two seats positioned back to back, or four seats around a central box. After about 1850 they were often more complex and were sometimes built around a central table or jardinière, or were of S-shape so that the sitters faced each other in the manner of the French tête-à-tête.

Ottomans were used a great deal in large spaces, such as ballrooms and galleries, where sitters were expected only to perch briefly for a rest, rather than lounge comfortably. Straight, one-sided versions called banquettes were also used in this way. The term ottoman also described pouffes and other fully upholstered stools; there was no visible woodwork. The name originated in the Near East. The ottoman required a great deal of upholstering with its deep buttoning and masses of fringes and both the back and seat were sprung.

75

and a matching gentleman's and lady's chair. A gentleman's chair was the same shape, but larger, than a lady's and had wooden arms with padded rests. Although rococo-style furniture of this type was still available even in the 1880s, generally by then the same basic types of furniture, but with straight contours and short turned legs and arm supports, were far more common.

In about 1860 another new piece of furniture appeared — the chesterfield. This was a fully upholstered sofa with horizontal arms and back of the same height, which were linked together in one continuous section. During the present century it has been fashionable to upholster a chesterfield in leather, but this would not have been done in the nineteenth century.

Yet another new piece of upholstery, the "cozy corner," became popular towards the end of the period. Composed of two high-backed, upholstered seats set at right angles to each other, these rather large objects were often combined with an arrangement of shelves for

A perfect vehicle for buttoning, the curvaceous Victorian ottoman (above) is a striking feature of this colorful and eclectic interior.

One of the enduringly popular types of sofa invented by the Victorians is the chesterfield (right). Comfortable either to sit or lie on, they have been widely used in all types of interior.

books or knickknacks and were positioned close to a window or near the fireplace to form a sort of inglenook.

By no means was all this furniture deep-buttoned. Shallow-buttoning was common too — although by 1830 proper buttons, not silk tufts, were used — and a great many pieces were not buttoned at all. Often only the back and the inside of the arms were buttoned; the seat remained smooth. There don't appear to have been any hard-and-fast rules about which types of furniture were buttoned and which were not, but for obvious reasons it appeared more frequently on well-stuffed, comfortable drawing-room furniture than on dining chairs.

Throughout the nineteenth century the furniture industry was still centered in London, the better firms such as Wright and Mansfield and Wilkinson's based in the West End, and the cheaper makers in the East End. Tottenham Court Road was a favorite shopping street for furniture and it was there that James Shoolbred & Co., a large manufacturer of upholstered furniture, had their retail showrooms.

· USE OF FABRICS ·

The fashion for elaborate French draperies, which had existed during the Regency period, persisted until the 1890s. Not only were win-

Conversational settees (above) were made in large numbers by the Victorians. It would be interesting to know how often they were actually used.

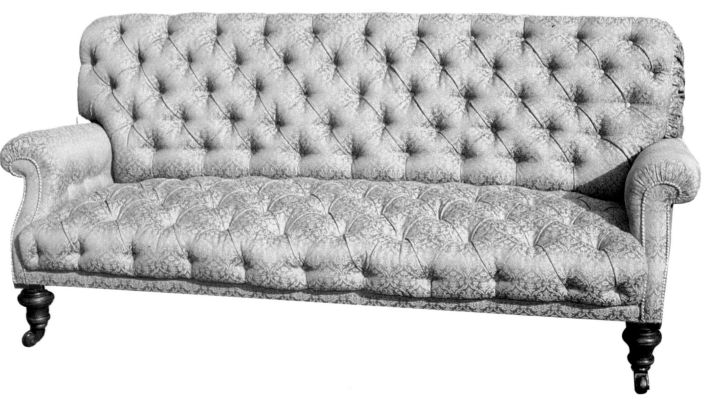

The extensive buttoning on this sofa (right) has been copied from a design for an identical sofa in a catalogue of the fashionable upholsterers Shoolbred & Co.

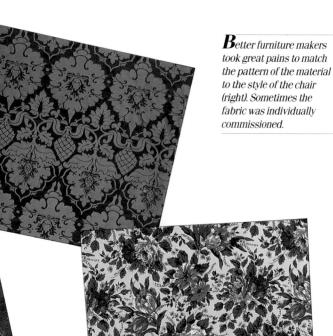

dows hung with elaborately shaped valances and vast folds of heavily fringed materials, but later in the century modified versions appeared over the doors and underneath the mantelpiece. Light and air were excluded from the room by thick lace curtains, and lace and crochet-bordered anti-macassars were used in abundance. These were strips of material hung over the backs of furniture to prevent the upholstery from being soiled by men's macassar hair oil. Enormous paisley shawls were casually draped over chairs and sofas and large numbers of cushions of varying size and pattern were randomly placed around the furniture.

The light and bright plain fabrics of the Regency period were replaced after about 1830 by heavier patterned materials such as velvet, brocatelle and worsted damasks. Later in the century plush was popular. A large number of silk fabrics was imported from France, a practice which until 1826 had been prohibited by law. The production of woven figured damasks with patterns ranging from Renaissance strapwork and gothic arches to naturalistic designs of fruit, birds and flowers, was made more cheaply and easily by the introduction of jacquard looms around 1830.

Not surprisingly, gothic-style chairs were usually upholstered with materials of gothic design and so forth, but large upholstered drawing-room pieces were mostly covered with fabrics in foliate patterns. Sometimes armchairs and sofas were upholstered in two different fabrics, one fabric was placed on the inside of the arms and the back and on the seat, and the other on the outside of the back and arms only.

Stripes were popular too, from about 1830 onwards on silk, and from about 1850 on damasks. On the latter, broad stripes were usually combined with bands of other ornament. As the century progressed, designs became increasingly eclectic and combinations on one fabric of classical columns, gothic arches and roses, for example, were not unknown.

Although patterned fabrics were by far the most common, plain materials, in particular plain velvet, were also used. Leather and hair-cloth were still popular for dining rooms and libraries and were usually bordered with matching gimp or tape and widely spaced brass nails, sometimes of a large size. Cheaper furniture might be covered with black American cloth, a thin but strong glazed cotton with a composite backing, which superficially resembled leather or horsehair.

Early in the period bright, deep colors such as scarlet and turquoise were fashionable, but by 1850 they had been ousted in favor of the darker colors such as the familiar maroon and bottle green which are considered so typical of Victorian taste. Patterned fabrics were usually still monochrome.

During the 1880s, when Moorish furnishings were at their most popular, oriental carpet, or cheap woven imitations, was used to upholster large, comfortable lounging pieces as well as smaller pieces of obviously oriental inspiration. It was often seen on folding chairs, where it was slung across the frame without any form of stuffing.

An exceptionally popular form of covering for smaller pieces, in particular footstools and

vesper chairs, was Berlin woolwork, a specific type of embroidery. A huge assortment of pre-colored designs — estimated in 1840 to have reached 14,000 — were directly imported from Berlin. Of course not all patterns were aimed at furniture, but those that were were largely floral; wreaths and bouquets of flowers worked in brilliant colors on a mostly black ground.

· TRIMMINGS ·

Trimmings made an important contribution to the overall design of all pieces and were generally much simpler than is thought to be the case. The majority of furniture was bordered with narrow, machine-made gimp of matching color, and deep upholstery frequently had edges and fabric joins decorated with variously colored twisted cord. On some early pieces, long, often floor-length fringes, with deep lattice-work headings, were fixed along the

*T*he most common Victorian trimmings (left) were narrow gimp and twisted cord. More elaborate fringes and tassels were regularly used on curtains and other drapery.

*T*hick tasselled fringes with deep headings (above) were considered particularly suitable for Moorish furnishings. Ottomans were suitable candidates for this treatment, whatever their covering.

*S*pecially shaped panels of decorative needlework were often set onto a piece with borders of velvet or other plain fabric (right). The join was disguised by cord.

*C*hintz was only occasionally used for fixed upholstery as on this exact copy of an 1841 chair (right).

79

*B*unches of full-blown roses on a white or cream ground were the most popular subject for cotton furnishing fabrics. This chintz (below) has recently been printed using 1850s hand blocks.

seat rail, having the dual purpose of providing a decorative feature and disguising the legs of the piece. Similar, but shorter, fringes were also used, sometimes on dining chairs, even when they were covered with leather. Tassels, which are thought to have been common, were sometimes attached to the facings of arms on sofas and easy chairs, but were more often seen on the corners of loose cushions. Heavy tasselled fringes were considered particularly suitable for Moorish divans and ottomans.

· LOOSE COVERS ·

A common practice, only recently recognized as such, was the use of chintz loose covers in reception rooms as well as bedrooms. The furnishing of the drawing room, where the fashionable hostess received her female visitors for morning calls and afternoon tea, was considered the province of the lady of the house, although she had to balance any inclinations towards feminine furnishings against the room's formal use for mixed gatherings in the evenings. Chintz covers provided an excellent compromise; pretty floral patterns gave the otherwise gloomy room a lighter feel, but could be quickly removed to show off the grander fabric underneath. They had other advantages as well; they protected the expensive fixed upholstery from dirt and wear and tear and were themselves easily washable. They were an ideal cover-up for worn-out upholstery, obviously a great boon to those with social aspirations but no money, and in the same vein could be used to modernize outdated pieces. In many houses

they were evidently used in the spring and summer months only.

Obviously loose covers varied in shape in as much as they had to follow the contours of the piece they were covering. What they did have in common was a deep, gathered or pleated floor-length skirt. Sometimes, but not always, they had piped edges. Occasionally chairs were seen with fixed chintz covers, but this seems to have been a rare occurrence, in drawing rooms at any rate.

The typical Victorian chintz had a pattern of large bunches of flowers, usually containing roses of some sort, although at other times subject matter included historical motifs, romantic pictorial scenes, exotic flora and fauna, and paisley and tartan designs. Even in its own time Victorian furniture was severely criticized by leading aesthetes and designers for its tasteless ornament and its reliance on historic styles. As early as 1860 reformist designers were advocating a return to simpler and, in their own words, more "honest" furniture, but their efforts were not to have much effect on the average household until the 1890s. Charles Eastlake in 1868 in his *Hints on Household Taste in furniture, upholstery and other details* spoke out against the purchase of furniture that wasn't functional and simple in design. His book went down well in England and the United States, where the bourgeoisie were eager for guidance on etiquette in all areas of their lives. However, clutter and overstuffed upholstery continued to be the dominant features.

ART AND CRAFTS

hough the majority were content to furnish their homes with factory-made furniture in historical styles, the movement for change was fuelled on a number of fronts by theorists, artists and designers who rejected the so-called commercial design. They formed groups and alliances in an attempt to return to what they saw as honest principles of design and "truth to materials." We tend to group them together as the Arts and Crafts movement, although their work and the theories behind them sprang from different sources. What they did have in common was a rejection of excessive carved ornament and commercialism.

· WILLIAM MORRIS ·

The Arts and Crafts movement is inseparable from the name of William Morris. This essentially romantic reformist movement had its origins in the writings of Pugin and John Ruskin, who wrote about the truth and purity of personal creation. In Ruskin's book *The Stones of Venice* (1851) there was a chapter "On the Nature of Gothic" which is said to have been the basis for the establishment of the Arts and Crafts movement.

80

Design Detail

The box cushions on the Sussex chair were shallow buttoned and without piping.

*D*espite the fact that the wallpaper, carpet and other textiles all have different patterns, this drawing room at Standen (above) presents a unified Morris interior.

*L*ater Morris & Co. furniture, like George Jack's Saville chair (left), differed little from those made by other manufacturers.

JOHN RUSKIN

Ruskin was not an architect or designer and yet he had a strong influence through his writings, which were firmly rooted in morality. Ruskin emphasized the need for doing away with division of labor so that the craftsman could take pleasure in his work and that the work should be for a purpose and in keeping with past styles.

Morris's greatest talent was as a textile designer and although his fabrics and wallpapers are now available in a wide variety of modern colorways (left), they are still instantly recognizable as his work. His experiments with natural dyes resulted in some wonderfully subtle and original color combinations.

The Morris Chair (below), designed by Morris's associate Philip Webb, was one of the most popular designs in the Sussex range of country chairs. It was based on a model made by an old Sussex carpenter, Ephraim Colman.

Morris founded his firm, Morris & Co., in 1861. Although not himself a furniture designer, Morris's company at first produced painted furniture with decoration romantically inspired by the Middle Ages. Soon they were reviving various forms of country furniture, some of which was upholstered in the traditional way using fabrics designed by Morris and woven or printed by craftsmen working under his supervision. His essentially two-dimensional textile and wallpaper designs were closely based on nature and mostly consisted of flowers and foliage, sometimes interspersed with birds or fruit, which flowed over the surface of the material.

According to his socialist principles, Morris had intended to produce "art for the people" but he soon discovered that hand-crafted furniture was too expensive for the average household. The firm's later catalogues make a clear distinction between "Cabinet Work," which was more commercial, and "Joiner-made and Cottage Furniture." Morris & Co.'s best-known chair is a bobbin-turned armchair with an adjustable back which was made with two large, loose, box-shaped cushions, without piping or any form of edging, and padded armrests. Morris's textile and wallpaper designs long outlived the popularity of his company's furniture.

NINETEENTH-CENTURY BRITAIN

uskin's writings and Morris's work inspired other groups of designers to move away from historic pastiche and over-ornamentation. One of these was the Century Guild, which was founded by the architect A.H. Mackmurdo in 1882. Others were the Art Workers' Guild of 1884, the Arts & Crafts Society, 1888, and the Guild of Handicraft, also 1888. In 1890 some of these designers together formed a business called Kenton & Co, through which they could sell their work. A number of them moved to Gloucestershire, which they felt was a more appropriate setting for hand-craftsmanship; they are usually referred to now as the Cotswold School.

While the designs of the various groups and individuals were often very different, they all had a common desire to reject the past and to take a fresh approach to design. Like the early work of Morris's firm, their furniture tended to be exclusive and had a limited market, but collectively they had a strong influence on commercial manufacturers and an identifiable style quickly evolved. At the time called "Quaint" and subsequently referred to as "English Art Nouveau," more recently it has been given the more apt title "Progressive."

· SEAT FURNITURE ·

Seat furniture was rather lightweight; chairs had very tall, rather narrow backs, and legs were thin and straight and without stretchers. Most were of square section and tapered near the foot before bulging out again to form a shallow block foot. The same shape, a sort of square mushroom, was sometimes seen on the top of the uprights as well.

On wooden-back chairs, decoration of the straight splat was minimal and often consisted of pierced hearts or spades, or shallow-carved, long-stemmed flowers. Sofas, perhaps better described as settles, were box-like with upholstered arms the same height as the back. The most progressive pieces were made in light oak or birch, but mahogany, beech or pine stained to look like mahogany, were also used.

Upholstery was almost completely flat, in strong contrast to the rounded sprung upholstery that was still widely available. Drop-in seats were common and sometimes the seat upholstery was set about an inch or so in from

Design Detail

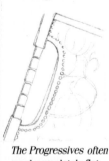

The Progressives often used completely flat upholstery, sometimes set back from the edge so the wooden frame appeared like a border.

*T*he swirling, naturalistic patterns of French art nouveau were openly condemned by British furniture designers as degenerate, yet the style was extensively adopted by leading textile manufacturers. Liberty's, whose fabrics were world famous, sold art nouveau materials by leading designers such as Voysey and Arthur Silver.

82

It is easy to see from the rather outlandish proportions of this mahogany chair (below) why there were so many objections to the art nouveau style. Finding a suitable fabric for such a distincive chair presents quite a problem, particularly as the back contains a decorative panel.

Wooden-panelled seating, built in to the structure of a room and fitted with loose cushions on the principle of a window seat, were an almost universal feature of fashionable decorative schemes such as this living room by Baillie Scott (right). They often included an unusually tall wing chair upholstered, like this one, with an art nouveau fabric.

the edges flush with the top of the seat frame. Narrow gimp to match the color of the fabric was the only form of trimming.

· NATURALISTIC DESIGNS ·

Although furniture itself had a strong vertical and horizontal emphasis, furniture fabrics were greatly influenced by the curvilinear designs of French art nouveau. Light-colored damasks woven with small, repetitive stylized plant forms were very popular, although more adventurous manufacturers used silk or cotton printed with larger and bolder patterns.

Some of the most distinctive fabrics were designed by Mackmurdo, their swirling naturalistic subjects inspired by the cloud and wave patterns seen on newly fashionable Japanese prints. Charles Voysey, a pupil of Mackmurdo, who in France was acknowledged as England's chief exponent of art nouveau, produced some popular designs for furnishing fabrics which were a cross between Mackmurdo's designs and the more formal bird and flower patterns of William Morris. Voysey was one of the many textile designers employed by Arthur Lazenby Liberty, whose avant-garde department store in London's Regent Street was famous throughout the world, particularly for its fabrics.

The art nouveau design of the back of this famous chair by A. H. Mackmurdo (right), designed for the Century Guild in 1882, was revolutionary for its date. Its rather dull, simple leather upholstery was in keeping with Arts and Crafts principles.

83

*T*horpe Lodge (right) was the home of Lord Montagu. He designed the furniture for the house (1904) in the style of the progressive furniture makers of the time.

*T*he fashionable art nouveau fabric that has been used to cover the elegant settee and armchair (below), made by Graham Morton of Stirling in the late 1890s, gives the unsubstantial-looking pieces some body.

Pale, pastel shades, particularly mauve, pink, green and grey, were the most popular colors for upholstery. They perfectly complemented the white or cream-colored woodwork, stencilled friezes and delicately patterned wallpapers, curtains and carpets seen in fashionable interiors. Rooms were not only much lighter and fresher, but were more sparsely furnished with far fewer ornaments than before. The clutter and oppression of the Victorian interior had disappeared.

THE SCOTTISH SCHOOL

Some of the most progressive furniture in Victorian Britain was produced by the Scottish architect Charles Rennie Mackintosh. His work is epitomized by his designs for Miss Cranston's Willow Tea Rooms in Glasgow where he was responsible for every item, from the murals on the walls to the crockery and teaspoons. It is quite evident from the chair designs that comfort was never a consideration, although obviously great trouble was taken with the design of the plain canvas upholstery with its stencilled, stylized rose motif.

Mackintosh was recognized abroad as an architect of originality; he influenced the work of the Vienna Secessionists in the early twentieth century. However, his output was small; he had few public commissions in Britain. He worked closely with his wife Margaret Macdonald, her sister Frances, and Frances's husband Herbert MacNair. They became known throughout Europe as the Glasgow Four.

Although upholstery played a small part in the designs, Mackintosh often designed alcoves for fitted upholstered seating or designed a chair to fit between cupboards. Squabs covered in his stencilled fabrics appeared on landings or window seats. The dining chairs usually had drop-in seats.

Upholstery played only a minor role in Mackintosh's striking and original interiors. His chairs fell into two groups (below). One was tall and designed to suit the proportions of the room (see right); the other was based on the cube.

Mackintosh's work is generally exemplified by Hill House, such as this light and airy guest bedroom (top). His furniture for the Glasgow School of Art building (left) was by contrast dark and rather heavy.

The table and barrel chairs (above) were designed for one of Miss Cranston's Glasgow tea rooms in about 1911. The semicircular drop-in seats are covered with leather.

EDWARDIAN REPRODUCTIONS

espite the efforts of progressive designers to move right away from historical styles, reproduction furniture was still popular in Britain in 1900. Renaissance, Elizabethan and Jacobean or Stuart furniture remained fashionable for dining rooms and libraries, while Queen Anne, Chippendale, Sheraton and *Painted Adams*, as well as eighteenth-century French furniture in both rococo and neo-classical styles, were bought for drawing rooms. The general trend for light rooms, sparsely furnished with rather spindly furniture, made *Adams* and Sheraton the most popular styles and white-painted *Adams* was used in bedrooms.

French furniture was seen in rather grand houses and was often of high quality. Manufacturers went to a lot of trouble to produce the right effects. Hindley & Wilkinson, one of the best makers, had their tapestry specially woven for them at Aubusson, in France, and the famous firm Gillows actually had their furniture made in France. English reproductions were not always so good and were often "in-the-style" rather than copies of earlier furniture. Many tended to have a mean, cheap appearance.

A great deal of this reproduction furniture was sold through the growing number of department stores, among them Harrods, Debenham & Freebody's and Maples.

· SUITES OF FURNITURE ·

Seat furniture was fashionably designed, but not necessarily sold, in sets. By 1890 lounging had become perfectly acceptable, even in company, and the three-piece suite was the most important feature of every drawing room regardless of social level. However, manufacturers were faced with the dilemma of attempting to re-create in eighteenth-century style the comfortable, deeply upholstered armchairs and sofas that people were now demanding. They mostly got around the problem by making basic sofas and chairs with the same contours — that is, rounded backs for Chippendale, square corners for Adam — and then upholstering them in the same fabric as the other, more closely reproduced, pieces. At their largest, drawing-room suites were composed of the standard three-piece, a number of single chairs with upholstered seats, similar "desk" versions with wooden arms, one or more high or low stools, a wing chair and a wooden-framed settee.

The better firms made some effort to reproduce the correct shape of the upholstery as well, although there was a tendency to use rather bland, pale, traditional-looking materials, in particular cotton damask woven with smallish floral patterns or stripes. Leather and velvet were used for the heavier, carved chairs in pre-eighteenth-century styles. Printed floral chintzes or cretonnes were still popular for loose covers, although they were sometimes used for fixed upholstery as well. A combination of the two, the loose covers being used just for the three-piece, was not uncommon.

86

Design Detail

Gathered and piped skirts on loose covers were popular at the beginning of this century.

The revival of eighteenth-century styles often resulted in a rather bland look. Suites of upholstery included a fully upholstered sofa and two armchairs as well as wood-frame chairs.

*P*ale, pastel shades were popular for Edwardian upholstery. This beautiful silk upholstery (right) has a heavy matching fringe and tassels stitched to the facings of the arms.

*G*athered borders on this pretty upholstery perfectly match the elegance of these "Sheraton" chairs (left). Designs for Sheraton-style furniture were produced by the designer A. Jonquet as early as 1880.

*A*round 1900 high-quality reproductions of Louis XV and XVI furniture became available. Its dainty upholstery (below) seems out of place with the Turkish rug and Saracenic table.

*C*hintz loose covers with a deep, gathered or pleated skirt had become the norm for three-piece suites in comfortable drawing rooms by 1900. This reproduction chair (below) has been covered in an appropriate Victorian floral chintz.

NINETEENTH-CENTURY STYLES

FRENCH INFLUENCE

·

ART NOUVEAU

·

JUGENDSTIL
AND THE SECESSION

·

DIVERSITY IN AMERICA

·

ARTS AND CRAFTS

*"Rococo" furniture in a
grand Palermo villa*

FRENCH INFLUENCE

n France the years following the Napoleonic wars saw a continuation of the social changes brought about by the Revolution. The bourgeoisie continued to grow in both numbers and prosperity and created new demands on the furnishing industry. To supply less extravagant furniture, French makers turned their attention towards cost-cutting industrialization, a move which was supported by the government in the form of industrial exhibitions held at the Louvre. Unfortunately, labor-saving machinery was often used to produce poor designs and the quality of French furniture made during the period of about 1815-50 is generally said to have been low.

Where design was concerned the same eclecticism was evident as that in England. The most popular styles were a heavier and less ornamented version of Empire, and "troubadour" or gothic, which was based on architecture of the late Middle Ages. Also fashionable was the practice of combining naturalistic carved details of sixteenth-century French Renaissance motifs such as classical figures with furniture of mostly Louis XVI type.

Until about 1840 all fashionable furniture was made in *bois clairs*, mostly light mahogany and bird's eye maple. After that date, dark-colored woods, such as ebony and stained pear and oak, became more popular and were sometimes inlaid with mother-of-pearl.

Until the middle of the century the majority of chairs followed Empire lines and had rounded backs and outward-curving saber legs. After about 1830 the legs sometimes had a distinctive inward-curving upper section above the saber curve. Empire-type scroll-ended *meridiennes*, or daybeds, with plain veneered wooden frames, and simple, fully upholstered sofas with slightly outward-curving arms, were both very fashionable and sometimes had an adjustable lower end.

The years 1850 to approximately 1890 saw France lose its status as a leader of world fashion in the decorative arts. The international exhibitions which were such important events in Europe's artistic calendar proved that although the quality of French craftsmanship was still superb, on the whole demand was for cheap and useful domestic furniture, in which artistic design generally played a small role.

Pastiches of past styles dominated interior

Design Detail

Nineteenth-century rococo upholstery was deep buttoned to give it a rounded, opulent shape.

*S*pecially woven tapestry was by far the most popular covering for antique-style chairs, particularly Louis XV pieces, with needlework as a cheaper alternative. Very grand classical pieces were covered with woven silks for which France's silk industry was still world famous.

*T*he elegant mahogany chair (left) is typical of the simple undecorated furniture demanded by the bourgeoisie after the Napoleonic wars.

90

design, although only native styles were reproduced. There were conventions for using individual styles in particular types of room — rococo and other Louis styles for drawing rooms, Renaissance for dining rooms and so on. Renaissance furniture was particularly popular after the Palais de Fontainebleau was refurbished in that style for Napoleon III.

· SHAPE AND VARIETY ·

In boudoirs as well as salons, drapery and fine upholstery were the most important features. The basic shape and finish of upholstery was similar to that in England at the time. Both shallow- and deep-buttoning were popular and trimmings took the form of narrow gimp or long, plain fringes, but without the deep, decorative headings. Comfort was a major consideration and stuffing was thick and well formed, particularly on sprung pieces.

The variety of seat furniture available was still much greater than elsewhere. All the traditional types of chairs, sofas and daybeds were reproduced alongside a number of ingenious new seats for two or three people on the lines of the eighteenth-century *confident*. Fully upholstered furniture was not so popular and dark woods or gilding were favored for carved frames. An ebonized finish was frequently seen on the large numbers of rather insubstantial-looking occasional chairs which became popular for informal use in salons. Sometimes gilded rather than ebonized, those chairs had either caned or upholstered seats. The shape of upholstery was related to their over-all style; if rococo, they were deep-buttoned and rounded, if Renaissance or Louis XVI, they were flat and square-edged.

Figured woven silks were the most common material for more expensive pieces, but cheaper ranges made great use of flowered cretonne rather than worsted fabrics. Tapestry was particularly popular for grander furniture and vast quantities of seat covers of various shapes were made at Aubusson and Beauvais. The re-upholstery of antique furniture in tapestry was so common that until recently it was assumed that tapestry had been the standard covering fabric in the eighteenth century as well.

Compared to English furnishings, for example, French furnishings, were less somber and colors were brighter. Although cotton loose covers were evidently used, they don't seem to have been so common.

As elsewhere, there were designers and artists who rejected industrialization and historical pastiches and worked towards a total break with the past where design was concerned. Their efforts culminated in the 1890s in the style *l'art nouveau*, which, if not universally admired, at least re-established France as the instigator of fashionable European design.

*G*ilded Louis XV furniture (right) had particular appeal for the newly prosperous bourgeoisie.

*D*eep-buttoning, heavy fringes and cord trimmings were used for upholstery in Europe as this German antler chair (left) shows.

ART NOUVEAU

rance had seen a long succession of revivals of historical styles during the nineteenth century and progressive designers were searching for a new style. This emerged during the 1890s with *l'art nouveau*. The name derived from the Paris shop which opened in 1895 and became a center for decorative art and furnishings, including imports from Asia. The new style swept through Europe and the United States, though it was most completely expressed in Belgium, where a great deal of architecture and interiors in this style can still be seen. The principal designers were Victor Horta and Henri van de Velde, who later worked in Germany.

Art nouveau was characterized by a completely new form of ornament based on plant forms. In general the furniture was rather elongated in form and decorated with flowing carving of twisted tendrils and flowers, undulating hair and trailing peacocks. The whiplash was a favorite motif. Although the basic framework was carefully balanced, the decoration was often asymmetrical. Inlays of colored woods and mother-of-pearl and decorative brass mounts were sometimes also used. The most important furniture designers in France were Louis Marjorelle and Hector Guimard.

One of the advantages of the style was that it could be satisfactorily applied to objects of virtually all types and sizes, making the creation of unified interiors — a concept which was very important — much easier. Upholstery was an integral part of the total design of each piece and, as such, didn't follow any regular pattern. Its depth and shape were determined by the contours of the chair. For many pieces the covering material had to be specially made. A number had leather upholstery decorated with stamped designs and edged with nails; others had plain silk covers decorated with stencilled, or printed stylized motifs and borders of gimp.

Although French art nouveau showed a totally fresh approach to design and is still thought by many to be a beautiful style, at the time it was generally thought by leading designers and the public in other countries to be degenerate. Because of its sinuous, flowing form, the art nouveau style was not suited to mass-production, and it was short-lived.

Design Detail

A close-up of a favorite art nouveau motif in its celebration of plant forms and nature.

The recent popularity of art nouveau furniture has been short-lived, as it was at the time of its inception. The importance of unity of design has made it difficult to integrate into other interiors.

*T*he design of this walnut chair with marquetry decoration (right) by Louis Majorelle is unusually angular for French art nouveau.

*L*eather was a popular covering for some art nouveau chairs. Stamped designs of swirling naturalistic patterns (far right) were similar to those on fabric covers.

*N*ot all art nouveau interiors were furnished with textiles of similar design and other styles were simultaneously used. This conventional fabric (right) is by the famous designer Fortuny.

*T*he individual shape of many art nouveau chairs and sofas demanded commissioned covers. The printed silk and long fringe on this Majorelle chair (right) has been carefully copied from the original.

*T*he most common subject for art nouveau textiles was the plant form. Stalks, tendrils and leaves could follow the line of any frame without looking unnatural (far right).

JUGENDSTIL AND THE SECESSION

n Germany and Austria art nouveau went through two phases. The first, called Jugendstil (after the magazine *Die Jugend*, in which art nouveau designs were first published in 1896), was based in Munich and displayed French characteristics, most notably the use of ornament derived from organic forms, although in a much more restrained way. Jugendstil was soon ousted in favor of the more architectural approach to modern design shown in the work of British designers, many of whom exhibited on the Continent.

Elsewhere in Germany and Austria, groups of designers were adopting the socialist principles of William Morris, but instead of making craft furniture, they turned their efforts towards allying good design with factory production, thereby putting good-quality furniture within the reach of the mass of the population. The two most important groups were the Wiener Werkstätte in Vienna and the Deutsche Werkstätte in Dresden. Exhibitions of "modern" furniture from all over Europe were regularly held by the Wiener Secession, an organization founded in 1897. It was at one of these exhibitions that the work of the Glasgow Four was so greatly admired.

The work of the Secession designers was much harsher and more geometric than elsewhere and the use of ornament was limited. Although by now quite different from French art nouveau, upholstery was surprisingly executed along similar lines.

94

Design Detail

The leather upholstery has a design stamped on it for decorative effect.

*L*ooking at this chair by Josef Hoffmann, it is clear that German and Austrian Secession designers admired the work of Charles Rennie Mackintosh. The furniture has the same clarity of line and strong vertical emphasis.

*T*he influence of French and Belgian art nouveau is evident in a small degree in this 1902 design (right) by Joseph Urban. Its flat brass-nailed leather upholstery gives life to an otherwise rather boring piece.

*T*he Victorian-style buttoned upholstery of this Hoffmann chair (below) looks rather out of place. The upholstery and the decorative side panels look more like a concession to public taste than a truly progressive feature.

95

*T*he simple armchair (above) with its equally simple upholstery can be seen as a source of inspiration for the tub armchairs invariably associated with the 1920s.

*T*he beautifully made walnut chair with simple mother-of-pearl inlay and brass feet (left) was designed by Otto Wagner in 1898. Wagner obviously had appearance rather than comfort in mind when he designed the completely flat leather seat.

DIVERSITY IN AMERICA

ashionable furniture in America was popularly made in predominantly historical styles. The pillar-and-scroll type of furniture, with flat veneered surfaces and elliptical scrolls, continued to be popular throughout the 1830s and 1840s. It was still, rather optimistically, referred to at the time as "Grecian," although the only obviously classical elements were the basic shapes of scroll-end sofas and monopodium pedestal tables. A large number of designs for furniture of this type appeared in 1840 in a Baltimore maker's pattern book, *The Cabinet-Maker's Assistant,* a publication which was very widely used. New York was by now fully established as the center of the furniture trade and it was from there that most new fashions emanated.

The gothic style was also fashionable between about 1830 and 1860, although less for furniture than for architecture. As in France, American gothic furniture was composed of architectural details, such as tracery, pointed arches and crockets, put together to form chairs and other furniture of essentially modern type. There was no attempt, as there was in England, to reproduce original forms. Gothic chairs were popularly covered in needlework, as were their contemporaries in Elizabethan style. So-called Elizabethan chairs were more closely based on English Restoration pieces and had tall backs, rather low seats and elaborate carving and spiral turnings.

Not surprisingly, French styles were also popular, particularly for drawing-room furniture. The first and most widespread style was rococo which, as elsewhere, was used for a large variety of upholstered sofas and *chaises longues,* and both easy and occasional chairs. However, the backs of many basic Louis XV forms were extensively decorated with heavy, pierced carving, which tended to give them a top-heavy look. The most popular woods for all seat furniture were mahogany and rosewood,

96

ADDING A FRINGE

A fringe can decorate a simple armrest but you can only use it if the arm is absolutely flat. It is attached in the same way as gimp (see page 39) and ideally should be glued and sewn or tacked.

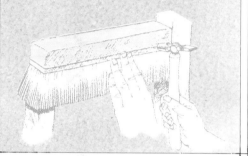

Design Detail

Extravagant borders of gathered fabric were applied all over furniture in nineteenth-century America.

A particularly extravagant and individual interpretation of the rococo style can be seen in the laminated and steam-bent furniture (right) of John Henry Belter. Bold choice of fabric is a good match for the slightly ostentatious design.

American middle-class homes were furnished in a similar manner to those in Europe. Rococo-style sprung upholstery was popular, as the balloon-backchairs and haircloth-covered easy chair in this painting (left) show.

NINETEENTH-CENTURY STYLES

although after about 1870 the grandest pieces were often ebonized or painted white and gilded.

· 1860-1900 ·

After about 1860 a form of Louis XVI furniture took over from rococo, although it was much heavier and gloomier than anything made in eighteenth-century France and relied more on neo-classical decorative detail than on basic construction. Chairs with rectangular or oval backs, broad, square seats and round, tapering legs were the most popular type and were made in mahogany or ebonized wood and decorated with gilded motifs.

During the second half of the century eclecticism and clutter were the most characteristic features of the middle-class home. French styles were joined by a rather ponderous form of French Renaissance and reproductions of earlier American furniture. There was a limited craze for "Saracenic" or Moorish furnishings, where upholstery played a predominant role. Fully upholstered divans were elaborately draped and edged with tassels and fringes, and piled with similarly decorated cushions. Oriental carpet, or velvet or plush woven in imitation of it, were considered suitable coverings.

A certain amount of furniture was made in the "early English" style which followed the principles, if not the actual designs, of Eastlake's *Hints on Household Taste*. Eastlake allied his work to mass production for the wider market. In practice his furniture represented a mixture of styles. Gothic construction and hinges were combined with low-relief carving of Japanese motifs, such as cherry blossom, chrysanthemums and cranes, and the turned spindles and brackets seen on contemporary English Queen Anne-style furniture. Chairs made in the Eastlake style were covered in leather with widely spaced brass nails covering a border of plain tape.

The shape of upholstery was considerably altered by the use of coiled springs from the 1820s onwards. Evidently springs were initially used because they were thought to provide a cheaper alternative to handstitched and stuffed upholstery, not because they could produce a more comfortable seat. At first they were attached at the bottom to wooden boards rather than webbing, and they were seldom held in place properly at the top, a practice which caused them to lean in towards the middle of the seat when the top cover was put on. As there was little additional stuffing around the edges, seats often had a domed appearance, a feature which did not appear in Europe. As better methods of attachment were devised, and it was realized that springs had to be combined with conventional stuffings, upholstery took on a smoother look.

Another distinctive development that was

These rather extraordinary chairs (left), made by Pottier & Stymus in New York in about 1875, were exhibited at the influential Philadelphia Centennial Exhibition. They are unique in their over-all design, but have delicate turned legs and carved decoration of classical inspiration. The marvellous turquoise and gold velvet upholstery has Renaissance patterns. The multi-colored fringe is typically nineteenth-century.

Nineteenth-century American makers devised various new forms of sofa, many with romantic associations. This small love seat (below) is prettily upholstered in a floral silk and has a typical matching machine-made gimp.

*P*attern books continued to be a popular source of design ideas. Designs in George Smith's 1826 book, from which this sofa (above) is copied, were particularly popular.

*T*he smooth, flat seat of this sofa (right), with its neatly piped edge, contrasts sharply with the skilful buttoning of the back.

CHAIRS WITHOUT JOINTS

In Austria a furniture maker in the Biedermeier style, Michael Thonet, developed a method of making chair frames without joints. He used the shipbuilding techniques of bending wood with steam and then laminated them into shape. In such a way the now familiar bentwood furniture was developed. Thonet introduced the rocking chair to Europe. His model had a leather-buttoned seat and back.

The best and most delicate rococo furniture in America was made by the German-born cabinet-maker John Henry Belter, who also patented a method of laminating wood which could be steamed and bent into any shape. This was a cheap way of mass-producing curved frames, although decoration had to be carved separately from solid timber.

widely adopted was the production of the Turkish frame chair, where the entire frame — apart from the legs, which could of course be seen — was made of metal. A small number of similar chairs was also produced in England.

· FABRICS AND COLORS ·

Industrialization of the furniture trade was advanced in the United States and makers were able to supply the growing demand for luxurious upholstery more easily than in Europe. In all homes elaborate drapery, especially curtains, played an important part. The choice of colors and materials depended to a large extent on the function, and consequently the style, of each room. In general the lighter, plainer materials of the Empire period were replaced in the middle of the century by heavier

fabrics in more somber colors, such as maroon, bottle and olive green and old gold. In the 1870s stronger colors such as red, blue, black, white and ivory were used.

Fabrics varied considerably during this period. During the second quarter of the century imported French patterned silks and velvets were used for the most luxurious furniture, while figured worsted fabrics, especially woolen damask, were widely used on less expensive pieces. By the middle of the century plush, brocatelle and patterned cotton velvets and damasks, often in two or more colors, were more common, as were printed cottons made in imitation of silk damask. During the second half of the century cotton chintzes and cheaper cretonnes were widely used for fixed as well as loose covers. While some patterns

were printed with decorative motifs from the various revival styles, the majority had lush floral patterns sometimes combined with broad stripes. Overall the full range of colors was employed.

American upholsterers seem to have been adventurous in their application of fabrics and made great use of pleating and gathering, particularly on fully upholstered pieces. Borders of gathered materials often ran along the front of the seat, up the facings of the arms and around the edge of the back. Cretonne covering usually had welted joins, while twisted cord was popular on sofas and easy chairs. Later, cretonne upholstery had a gathered or fluted flounce or skirt covering the legs.

Decorative slip covers were still supplied for most upholstery and reached right to the ground, even on stools and occasional chairs. A Boston upholsterer, writing in 1890, advised that slip covers were essential for the "dusty season" when even the wooden parts of upholstered furniture would benefit from protection. This implies that the seasonal use of loose covers in England, France and elsewhere was also primarily practical and was not related to any considerations of appearance.

ARTS AND CRAFTS

Arts and Crafts influence wasn't felt much in America until the 1890s, when small groups of craftsmen began producing simple, handmade furniture according to Morris's principles. The most successful designs were produced by the Roycroft Community set up by Elbert Hubbard in 1895, and Gustav Stickley's Craftsman Workshop, both in New York. They produced simple, rectilinear oak furniture, upholstered, if at all, in plain leather. Stickley's designs were reproduced in his own magazine, *The Craftsman,* which he began in 1901 after a trip to Europe. The magazine contained features about furniture design and socialism and did a great deal to promote Arts and Crafts ideals among furniture makers.

By the end of the nineteenth century the status of the upholsterer was extremely low; although upholstery had been the most prominent feature of drawing rooms for several decades, the upholsterer had been considered very much a tradesman rather than a craftsman. For example, some proponents of the Arts and Crafts movement, whose principal aim was to return to hand-craftsmanship, advocated the use of "factory units" for upholstery. These were machine-made metal frames supporting coiled springs and borders of steel wire which came ready-covered with material.

Surprisingly art nouveau furniture was never very popular in America, although Louis Tiffany's *favrile* glass was internationally famous. One maker who did adopt the style successfully was Charles Rohlfs from Buffalo, but his work was all privately commissioned and had little effect on commercial producers.

The most progressive designer was the architect Frank Lloyd Wright, who inherited from his employer, Louis Sullivan, an interest in simple, geometric forms. He is usually associated with furniture made between the wars, although he was designing astonishingly modern pieces in the early 1900s. Like so much aggressively modern furniture, upholstery played a small role in the overall design.

The American furniture maker Gustav Stickley was greatly influenced by the work of William Morris and others during a visit to Britain. He made simple oak furniture along Arts and Crafts lines, and edited an influential magazine, The Craftsman, *which promoted Arts and Crafts ideas.*

One of the most progressive American designers around the turn of the century was Frank Lloyd Wright. Although he is usually associated with a later period, some of his early work was in a straightforward Arts and Crafts style (left).

An oak reclining chair (left) by Gustav Stickley was inspired by Philip Webb's Morris & Co. adjustable-back armchair.

Charles Rohlfs, who opened a workshop in Buffalo in 1890, was one of the few American furniture makers who adopted the French art nouveau as well as the Arts and Crafts style (right). He was widely recognized as a progressive designer on the Continent.

This armchair (above) by the maker B. Harley dates from around 1900. The typically plain Arts and Crafts upholstery is subordinate to the over-all shape.

BETWEEN THE WARS

THE BAUHAUS
·
ART DECO
·
REVIVALISM IN THE USA
·
ORGANIC FURNITURE
·
REVIVALISM AND MODERNISM

Salvador Dali's Mae West
Lips sofa, 1936

THE BAUHAUS

he period between the two world wars saw the emergence of an International Modern style. The most significant design developments occurred at the Bauhaus, a school of design founded in Germany in 1919 by the architect Walter Gropius. Its basic philosophy was the eradication of all historical concepts of design and the creation of a totally free approach in which function played the major role. Unity of design within a building,

the use of new materials and the suitability of objects for mass production by machine were other principles which the Bauhaus advocated.

The most important Bauhaus designers were Marcel Breuer and Mies van der Rohe; both experimented successfully with tubular steel as a suitable material for furniture. Its use is said to have been inspired by the metal handlebars of Breuer's bicycle, but whatever its origin, metal has since become a universally popular medium for chair frames, particularly those

which can be stacked, and many original Bauhaus designs are still in production today.

The majority of Bauhaus chairs cannot really be described as upholstered; seats, backs and arms were composed of strips of fabric — mostly leather or canvas — simply slung across

104

Marcel Breuer's Wassily chair, made of tubular steel with leather strips slung across the frame, was designed at the Bauhaus in the 1920s. It is still popular today, particularly for office interiors. In domestic settings tubular steel has never blended easily with more decorative furnishings.

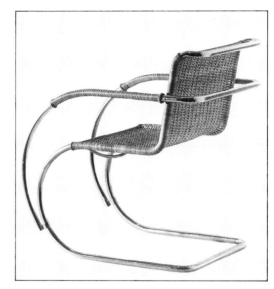

The cantilever chair (above) was originally designed by Mies van der Rohe. The use of woven cane was a departure from the usual leather or canvas.

Mies van der Rohe's famous steel and leather Barcelona chair (left) was designed in 1929. It was one of the few Bauhaus designs that used padded upholstery.

The first tubular steel chair (below) was produced by the de Stijl designer Mart Stam in 1924. The tubing was not continuous.

the frame. A few had woven cane seats and backs. A number of Mies's pieces, most notably his famous Barcelona chair, had shallow-buttoned and piped leather box cushions supported by leather straps. The Barcelona chair was made for the German pavilion at the Barcelona exhibition in 1929. Despite being one of the most famous examples of modern furniture, it is difficult to produce and uncomfortable.

· TEXTILES ·

Textiles suitable for upholstery were produced in the Bauhaus weaving workshop, where, surprisingly, handweaving was considered preferable to machine methods. Despite this apparently backward-looking attitude, experiments were made with new dyeing techniques and the production of light-reflecting and sound-absorbing materials, and some of the ideas which emanated from the workshop were subsequently adopted by large textile manufacturers. Ironically, many of the predominantly abstract Bauhaus textiles had a naive, almost ethnic look, very different from the stark modern furniture and objects that they accompanied.

Although universally admired by leading designers and aesthetes in many parts of the world, in their own time the products of the Bauhaus were not popular with the average householder, Nor was the Bauhaus itself popular with the German government. Condemned as decadent and subversive, it was closed down by the Nazis in 1933 and the majority of its designers emigrated to America where they were able to continue their teaching in a more receptive atmosphere.

At the *Exposition des Arts Decoratifs et Industriels* in Paris in 1925, the exhibits introduced a new style in design; this has become known as Art Deco, also called Jazz Moderne after some of the motifs which were inspired by Cubist and Futurist paintings. A condition of entry to the exhibition was that the exhibit should be entirely contemporary and not revive any past styles.

However, the furniture was fairly traditional in type but modern in its mostly two-dimensional and geometric surface ornament. The most common decorative features were stylized flowers, in particular roses. The best pieces were made in luxurious materials, such as lacquer, ebony, ivory, shagreen, sharkskin and highly figured woods. Brilliantly colored, patterned fabrics, many with an oriental flavor inspired by the flamboyant designs of Leon Bakst for Diaghilev's *Ballets Russes*, and plain-colored satin and various furs and animal skins were used for upholstery. Armchairs and sofas had semi-circular backs and square or rectangular bases with deep, smoothly rounded upholstery and large, soft seat cushions. Chairs of gondola form were particularly fashionable

The rounded back and tub shape of this French armchair (right) are counter-balanced by the straight sides and seat. The stylized flowers on the seat and back are instantly recognizable as Jazz Moderne.

The bright colors and exotic subject matter on this tapestry-covered suite by Maurice Dufrêne (right) reflect the influence of the Ballets Russes.

UPHOLSTERY STYLES

LE CORBUSIER

In France the most avant-garde designers were Le Corbusier, his associate Charlotte Perriand, and his cousin, Pierre Jeanneret. They preferred to see furniture as equipment, which, although contributing to the over-all design of the building, was essentially there to be used. Like the Bauhaus designs, Le Corbusier's furniture was radically different from the fashionable Parisian furniture with which it was contemporary. The Grand Confort, rather like a steel cage holding the large cushions in place, was Le Corbusier's most famous piece of furniture. It established a new idea, that the frame of the piece of furniture was "outside" and not "within" the upholstery.

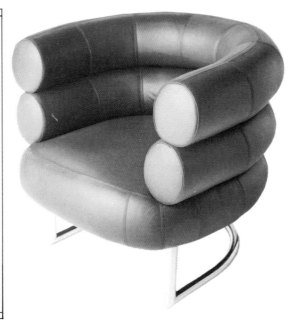

***T**he British-born designer Eileen Gray, who spent almost all her working life in Paris, was already producing some extraordinarily progressive and individual furniture in the 1920s. The tubular armchair (above) is upholstered in leather.*

***T**he Grand Confort chair (above) designed by Le Corbusier is made from chromed steel and leather. Despite its square upholstery, it is extremely comfortable.*

***E**mile-Jacques Ruhlmann favored plain satin upholstery in pale, muted colors to show his choice of figured veneers and other luxurious materials to advantage. This chair (right) is made from macassar ebony.*

for dining as well as bed and drawing rooms.

The most influential furniture designers were Paul Follot and Emile-Jacques Ruhlmann. (Follot was later to work for the furniture house Waring & Gillow in London in an attempt to introduce the better aspects of modern taste to the British.) Their highly individualistic furniture was available from the furnishing sections of Parisian department stores, such as *Bon Marché* and *Au Printemps*.

· REVIVALISM ·

Naturally not everyone could afford such exotic furniture and for the majority of the population period revivals continued to be fashionable. Indeed, during the 1930s even the most influential interior designers were combining antiques — or antique-style furniture and objets d'art – with new pieces in otherwise modern settings. The most popular styles for this rather élitist form of furnishing were those which had not been fashionable for some time – Biedermeier, Victorian and Ancient Greek – rather than the more conventional Louis styles. Both classical and primitive sculpture was also particularly fashionable, as were, for a short time, blackamoors.

· THE INTERIOR DESIGN ·

This was a time when rooms became more open plan, with areas set aside for specific activities, such as dining. Both walls and floors were popularly plain in color, frequently black or white, often yellow. Sometimes one wall was covered with mirror glass. Plain-colored satin, mostly white or beige and to a lesser extent grey, mauve or pink, continued to be the favorite material for upholstery, although leather and fur were still used as well. Animal furs and thick-piled, geometric-patterned wool rugs were fashionable on floors. Curtains were mostly plain or with an unobtrusive pattern with a vertical bias. They were usually hung straight to the ground, without any form of valance. If they were draped, it was done loosely in the antique Greek way.

For living rooms particularly the shape of upholstery tended to become more angular, although rounded backs remained common. The influence of Greek *klismos* chairs is evident in the shape of legs which were square in section and had a gentle outward curve. The upholstery itself remained smooth and edges were defined by simple piping. In the traditional French manner upholstery was set within, rather than over, the wooden frame, even on sofas and deep armchairs. In bedrooms, deep-buttoning and quilting were popular for chairs, stools and the head and base boards of beds.

REVIVALISM IN THE USA

t is significant that no American exhibits appeared at the Paris Exposition in 1925. With the exception of the work of Frank Lloyd Wright, there had been few notable advances in American furniture design. Far greater attention was paid to architecture, in particular to the development of the skyscraper. It is interesting to see that while buildings themselves were being decorated with a whole new range of Art Deco motifs such as sunbursts, stylized birds, leaping deer, bolts of lightning, ziggurats and other geometric designs, their interiors were largely fitted with furniture of traditional type. This is particularly surprising considering the importance attached by contemporary architects elsewhere to total unity of design within a building, a concept which Wright took to its extreme by designing individual furniture reflecting the shape of specific rooms, thereby making it quite unsuitable for use in any other interior.

Although the exotic materials and decorative repertoire of the French exhibits at the Paris Exhibition in 1925 began to appear occasionally on some American furniture after that date, the Wall Street crash effectively removed the majority of potential clients for this exclusive, and therefore rather expensive, type of furniture.

· OVERSEAS INFLUENCE ·

Throughout the 1920s and 1930s the dominant theme of virtually all domestic interiors was period revivalism. Fashionable styles included all those which had previously been popular plus Hispanic furnishings and reproductions of early American pieces. Both furniture and objets d'art were acquired more for their decorative qualities than for their standard of workmanship or, indeed, their suitability for the room in which they were housed.

Particularly popular in fashionable New York were English eighteenth-century styles, a trend which was due largely to the influence of America's leading designer, Elsie de Wolfe, who could be described as the first interior designer. She had her greatest success in the first decade of the century. It is to her that the popularity of floral-patterned chintz loose covers is usually attributed. Otherwise upholstery on revival furniture was little changed from the preceding period. Appropriately patterned silks, velvets and damasks were still in use.

During the 1930s there was a tendency to up-date period interiors in the same manner as in France by providing a modern setting for more decorative antiques. Walls were fashionably white, beige or cream and carpets, curtains and upholstery were also plain and in muted colors. "All-white" interiors were fashionable here too, inspired by Syrie Maugham, particularly in Hollywood, where fantastically well-paid film stars could improve their image by entertaining in the most up-to-date interiors.

As the decade progressed, "Modernism" began to take a stronger hold and the streamlined look of International Style architecture was adapted to furniture. Although initially mostly confined to offices, bars, restaurants, trains and ocean liners, modern chromium-plated, tubular steel furniture with glass or black lacquered surfaces was also seen in fashionable houses and apartments, particularly in all-white settings. Unfortunately, because this type of furniture was quite cheap to make, much of it was of rather dubious quality.

A fashionable New York apartment in the 1930s (above) has an unpatterned carpet and curtains, and plain white walls emphasizing the decorative qualities of the furniture.

Design Detail

Floral chintz loose covers were one of the features of Elsie de Wolfe's interiors in the early years of the twentieth century.

*T*he most progressive designer in America between the wars was Frank Lloyd Wright. This quite extraordinary enamelled steel chair with walnut arms (right) was made en suite with an equally unusual desk for Herbert Fisk Johnson in the Johnson Wax Building, Wisconsin, in the late 1930s.

108

*B*ecause all Wright's work was individually commissioned, and each piece of furniture was designed to reflect the shape of the room in which it stood, out of context some of his designs (right) often look awkward, if not bizarre.

*A*lthough not a particularly distinctive material, the cream and yellow striped covering on the seat furniture in this office suite by Wright (above), designed for the Edgar Kauffman department store in Pittsburgh in 1937, was obviously chosen with care. Textured fabrics, mostly with a wool base, became increasingly popular for upholstery during the 1930s.

109

*T*he tubular steel furniture in this 1930s apartment (right) looks rather cheap in the otherwise sophisticated "modernist" interior, particularly when compared with the solidly built and deeply upholstered wood-framed sofa.

ORGANIC FURNITURE

During the 1930s Bauhaus principles of modern design were successfully adopted in Scandinavia, but instead of metal, designers continued to use wood, the more traditional and plentiful material. Experiments with wood laminates of various kinds, and with the sort of molding techniques used in Austria in the nineteenth century by Michael Thonet for his bentwood furniture, combined to produce some simple "organic" seat furniture, closely related to the shape of the human body and admirably suited to factory production. Seats and backs were often formed from one continuous piece of laminated wood; upholstery, if present at all, was insignificant. Plain woven materials, in particular wool, appeared most frequently, and black or charcoal grey, which contrasted sharply with the light-colored wood frames, were the most fashionable colors.

Equally modern in its design, but less clinical in its appearance than its German metal counterparts, Scandinavian furniture was subsequently marketed with great success in Britain and America. In Scandinavia the firm *Futura*, and in England *Finmar*, did much to spread its popularity. The most important designers were Bruno Mathesson in Sweden and Alvar Aalto in Finland. During his two-year stay in Britain in the early thirties, Marcel Breuer designed some laminated furniture of Scandinavian inspiration for the *Isokon Furniture Company* in London.

Design Detail

Laminated stacking chairs were a successful example of organic furniture by Alvar Aalto.

In terms of cheapness and technical possibilities, laminated wood was little different from tubular steel and had the added advantage of being more acceptable to conservative-minded buyers. This daybed was one of several plywood pieces designed by Marcel Breuer in the early thirties. Its loose upholstered mattress is obviously well used.

110

MAKING PIPING

To make piping, the fabric has to be cut on the bias. Find the diagonal by folding the fabric into a triangle and cut the first piece of bias along the foldline. Measure and mark the remaining widths parallel to the first. Cut the required length. When the lengths are joined, make sure that the weave goes in the same direction. With right sides together, pin the strips at a 90° angle and machine stitch across them at a 45° angle.

Press the seams open. Fold the bias strip over the piping cord and pin in place. Using the zipper foot on the machine, and a thread the same color as the fabric, sew the folds together close to the cord. Piping cord comes in several thicknesses and if you are replacing old piping, which is often the first fabric to suffer from wear and tear, use an old piece as a guide to the size.

REPLACING PIPING

Keep the fabric on the piece of furniture in place with special skewers and pull out the old piping. Remove it and set in the new piping. Ladder stitch it in place. This is basically a backstitch where a curved needle is passed through all four thicknesses and brought out just behind the point where it was inserted. This stitch is used by upholsterers to join fabric invisibly. Tighten the thread every five or six stitches.

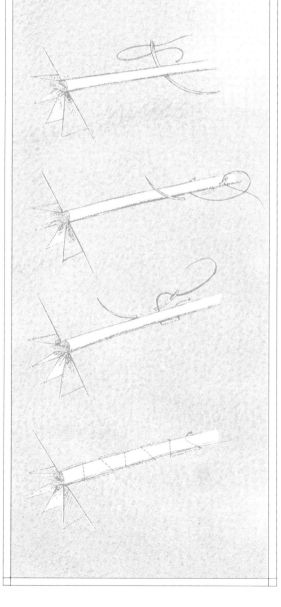

The laminated armchair (above) designed by Alvar Aalto in 1936 was a popular form in its time. Although the Scandinavian designers generally preferred plain, dark fabrics for upholstery at this date, virtually any color or pattern would look good on this simple frame.

Although this woven fabric (right) looks modern, it was designed in 1922 by a Bauhaus student and is therefore appropriate for a chair of this date. Its textured surface, which breaks up the stripe, anticipated the popularity of textured folkweave fabrics in the late thirties.

111

REVIVALISM AND MODERNISM

The present-day image of Britain in the 1920s is one of colorful modernity, outrageous fashions, racy cars, glitter and gaiety. In reality the situation was very different. While the shrinking upper classes continued to spend their family fortunes on new and streamlined consumer goods, the majority of the population lived under severe economic restraints and there was little money to spare for spending on new home furnishings and, consequently, little demand for new and adventurous styles. Until the end of the decade popular taste remained extremely conservative. Reproduction furniture, particularly that in Georgian and Jacobean styles — the latter perfectly suited to the rows of suburban mock-Tudor houses spreading outwards from the cities — continued to dominate the furniture industry.

The industry itself, like all other industries, was suffering from a shortage of labor, and investors were reluctant to put money into any new ventures if a suitable return could not be guaranteed. These were hardly promising circumstances for the introduction of new styles and ideas.

· POPULAR INTERIORS ·

As they had before the war, the department stores provided the greatest range of furnishings and most had a design studio which advised on interior decoration. Fashionable houses were fitted with rooms in one of two styles: Tudor interiors had oak panelling, Elizabethan-style plaster friezes and ceilings, and lead-paned windows. Fireplaces had brick or stone hearths. In dining rooms, trestles or rather clumsy extending tables with bulbous turned legs were accompanied by Cromwellian chairs covered with velvet or woven "tapestry" covers. Knole sofas were particularly popular and were mostly covered in velvet or conventional damask; red was the most frequent color for all materials.

In Georgian interiors, natural, polished pine panelling and parquet floors scattered with rugs were popular. Mahogany furniture, mostly in Queen Anne style, was covered in damask, tapestry, brocade or brocatelle. Wing chairs, upholstered to match the ubiquitous and suitably shaped three-piece suite, were fashion-

This re-creation of a fashionable 1930s drawing room in a suburban house is evocative of pre-war respectable modernity. The tiled fireplace, laminated plywood occasional tables, Art Deco mirror and light fittings and parchment lampshade, are perfectly in tune with the large and rather clumsy three-piece suite with its abstract-patterned moquette covers.

able. Georgian interiors often included cabinets, long-case clocks and occasional tables decorated in imitation of Chinese lacquer. Fabrics patterned with Chinese designs on a black background were particularly sought after. This led to a limited fashion for lacquered *bergère* armchairs and sofas which had lacquered frames, caned backs and arms and deeply stuffed, loose seat cushions with piped edges. Although by far the majority of lacquer had a black ground, red was also fashionable.

· REPRODUCTION FURNITURE ·

A great deal of repro furniture of this date was of dubious quality. Stamped ornament on Tudor pieces and glossy French-polished veneers on cheap plywood construction were common.

The general preference for reproduction furniture did not, of course, prevent the development of modernistic styles. Even before the war, furniture on modern lines — of simple rectilinear form with very little, if any, surface ornament — was being sold at Heal & Son in London's Tottenham Court Road. Much of it was designed by Ambrose Heal himself and although at the time it was considered progressive, it was in fact totally traditional in form and largely inspired by the teachings of William Morris, despite its intended manufacture by machine.

Ambrose Heal was one of the first British designers to attempt to ally good design with industrial production. He was a member of the Design & Industries Association founded in 1915 to establish a better standard of industrial design and to educate the public through exhibitions of contemporary products.

Initially Heal's "traditional-modern" furniture designs were largely restricted to bedroom and other carcase furniture, and Heal's catalogues of the 1920s show mostly chairs, sofas and divans of traditional Victorian form. Many of the deep-buttoned chesterfields and *chaises longues* have no doubt since been sold as such.

· MATERIALS ·

In 1920 recommended fabrics for curtains, loose covers and upholstery included printed linen, chintzes and cretonnes and plain and printed silks. Patterns were mostly floral or chinoiserie and dining chairs were usually covered in corduroy or plain or checked cotton.

HOME UPHOLSTERY

211—Puffed cushion—making the puffings. 212—Piping the mat for the puffed satin cushion. 213—Puffed cushion with top made from a Chinese mat. 214—Two decorative cushions—a satin bolster and a round one in puffed velvet.

The colors and general decoration of this Ambrose Heal design for a sitting room (right) give it a modern feel. Although simple, the furniture itself is traditional. The Sheraton-style table, Chippendale wine table and the pair of deeply upholstered chesterfields with Sheraton legs had been around for 20 years before this setting was contrived.

At a time when even wealthy families had to curtail their household spending, there was great demand for do-it-yourself manuals (below). They contained, for example, advice on how to produce luxurious French-style cushions to brighten up your drawing room and how to disguise your old and faded upholstery with bright and fresh loose covers.

HOME UPHOLSTERY

228—Cretonne for loose cover chalked and partly cut out. 229—How the back opening of the cover is arranged. 230—The finished cover—tailored style. 231—The finished cover, with frill and small loose cushion.

Throughout the 1920s Heal's sold a cheap range of cottage-quality furniture of which the most distinctive and popular item was a three-piece suite covered in brown hide, edged with brass, with matching velveteen seat cushions.

Scandinavian-influenced armchairs, sofas and settees were made with plain, wooden frames, with either rounded or square arms and straight square-sectioned legs. Both fixed and loose upholstery was smooth in shape and without trimming, apart from occasional piping. Jazzy-patterned moquette (a sort of velvet), woven cotton tapestry, repp, shot canvas and silks were popular materials, their abstract designs inspired by contemporary paintings. Floral as well as geometric patterns and waves, zigzags, checks, and stripes were fashionable but had a strictly two-dimensional appearance. At the end of the thirties coarse-textured "folk-weave" fabrics were coming into vogue.

The same types of fabrics, but with conventional patterns, as well as the traditional damasks, were used on reproduction pieces, for

113

*A*ll the right elements are present in this late thirties room (above) — plain surfaces, a glass table, a bear skin rug, a jazzy modern armchair and textured fabrics.

*T*his dining suite (right) is probably the closest British tubular steel furniture manufacturers ever got to the sophistication of the Bauhaus designers.

114

*T*his wide range of 1930s catalogue chairs (below) just missed the boat when compared with their French counterparts. The rather dowdy-looking fabrics don't help.

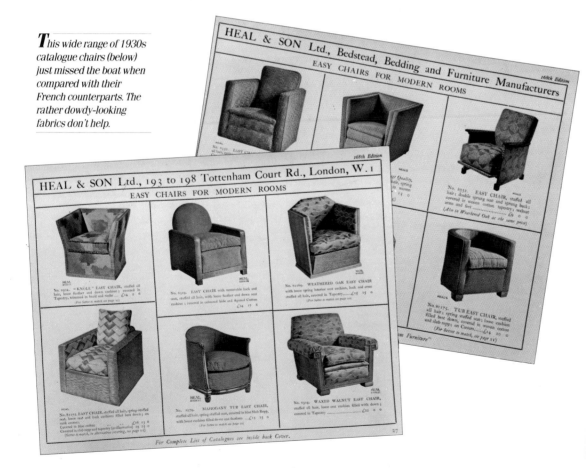

BRITISH MODERNISTIC STYLE

The last years of the decade saw the emergence of the first truly modernistic British furniture in the form of simple, rectilinear pieces on the lines of Heal, combined with the glossy veneered finish favored by the repro makers. For decoration this type of furniture relied mostly on the figuring of the wood — walnut, amboyna, rosewood and other exotic timbers were favored — but handles and simple lines of inlay executed in mother-of-pearl, ivory and ebony also provided some form of ornament. This type of furniture was not dissimilar to contemporary French furniture, many examples of which were optimistically displayed to the conservative British public at exhibitions at Shoolbred's in 1928 and Waring's in 1929. It was, however, inevitably expensive, and most manufacturers began to produce simple ranges of cheaper pieces in weathered or fumed oak or laminated beech along the lines of Scandinavian furniture.

which there was still a sizeable market. Heal's sold a special range of "old-fashioned fabrics" to "harmonize with old furniture" which included heavy cotton tapestry and printed linens and chintzes with patterns based on old embroideries and other textiles. Fabrics of all types were available in a full range of shades from pastels to fall browns, reds and yellows, with cream as one of the most popular background colors.

The same types of modern materials, including leather, were used for upholstery on chromium-plated tubular steel furniture, first marketed in Britain in the late twenties. Although it was believed to be the perfect choice for anyone with truly modernistic taste, it was never really popular for domestic use and was mostly seen in cafes and offices. The largest British manufacturers were PEL (Practical Equipment Limited) and Cox & Co.

To offset modern furniture to advantage it was considered advisable to have plain walls, either white or cream in color. Lightly textured wallpapers with narrow borders of geometric or floral patterns were particularly popular and were accompanied by parquet floors and small abstract-patterned rugs.

While the majority of middle-class homes were still furnished from the department stores, the most sophisticated interiors of the thirties were created by a growing band of fashionable independent interior decorators. One of the most successful of these was Syrie Maugham, whose name is commonly linked with the fashion for all-white rooms. Like most of her contemporaries abroad, Syrie Maugham was an advocate of period revivalism, and decorative antiques dating mostly from the Regency and Victorian periods were combined with plain walls and floors, simply draped curtains and modern fabrics and rugs. Furs of various kinds were used for upholstery as well as floor coverings. In bedrooms plain satin was a great favorite and was often quilted.

In interiors of all types, fully upholstered deep armchairs and sofas continued to be popular. Mostly of very large proportions, they tended to have horizontal backs and broad arms that were slightly higher at the front than at the back. Satin — just about the most impractical fabric you could choose for a three-piece suite — was thought to be the most sophisticated material, but woven, textured fabrics and cotton covers were more common for pieces in everyday use. Brown, or black, and white, lemon yellow and lime green were fashionable colors and cyclamen pink was often seen in bedrooms.

115

Streamlined architecture and decorative French-style furniture were fashionably combined in public buildings. In the main lounge on the Queen Mary *in 1935 (right) the carpet appears to clash with the upholstery fabrics.*

The glass and zebra skin armchair (above) was devised by the interior designer Denham Maclaren for his own use in 1930. No doubt a client would have been delighted with his efforts.

POST-WAR TRENDS

NEW TECHNOLOGY
·
INNOVATION IN ITALY
·
UTILITY FURNITURE
·
QUALITY OF DESIGN

*A stylish living room in the
1950s*

NEW TECHNOLOGY

After the war, while Europe struggled to re-establish its traditional industries, in America, manufacturers and designers were developing new materials and technology.

Since the 1890s the progressive architect and designer Frank Lloyd Wright had been designing the interiors of his buildings. His furniture was described in the 1953 catalogue of the Museum of Modern Art as "blocky," not a word often used in furniture description, but it does vividly describe 1940s and 1950s domestic upholstered furniture.

The most successful furniture designers of this period were the architect/designers Charles Eames and Eero Saarinen. They were helped by the patronage and manufacturing facilities of two established companies: the Herman Miller Furniture Company and Knoll International. Their first experiments were with metal and wood laminates and were seen at the Museum of Modern Art in 1940. The Museum's new department of industrial design held a competition which Eames and Saarinen won.

Within the next 10 years, their award-winning chair was produced in fiberglass and plastic all molded in a single piece. The shell was covered with a thin layer of rubber or foam, then a top fabric, and the whole thing attached to metal or wooden legs. Most of the chairs could be dismantled for storage and transport.

This furniture was designed to be comfortable, to mold the body, and in many cases to be interchangeable, or at least complementary, between the home and the office. It was not intended to last long.

Those that followed in this same fluid form were all named by their designers with descriptive labels. The "Womb" chair by Eero Saarinen (1945) was accompanied by a foot stool for those who wanted to take their shoes off and curl up in the chair. The "Tulip" table and chair, also by Saarinen (1957), sat on a circular base and metal pedestals, lacquered in white, and the chair had woolen upholstery.

· THE MODERN ARMCHAIR ·

It was Charles Eames who produced a modern design for the armchair. Molded plywood still provided the shell, but the cushions were leather, an ottoman accompanied it and the metal base allowed the chair to swivel. Saarinen's womb chair also provided a degree of comfort for those who rejected the old stuffed sofa.

The most expensive upholstered furniture

Charles Eames' swivel chair and ottoman (left) has an ageless quality. More than 30 years since its inception, it still has a contemporary look.

This design for a chair of lounging shape (above) is reminiscent of the fifties, but was actually designed by Charles Eames and Eero Saarinen in 1940.

was covered with leather, usually black. Plain woven fabrics in natural colors were popular; they were similar to those woven in the Bauhaus weaving shop in the 1920s. Surprisingly the large textile manufacturers were rather slow to gear the production of synthetic fibers used for industrial textiles towards the upholstery trade, and it was only a small number of progressive interior designers who experimented with the utilitarian materials. Trimmings were virtually non-existent.

Most sofas and upholstered chairs at this time were made up of square or rectangular blocks of foam attached to a metal or wooden frame. Occasionally they were shallow-buttoned. The blocky appearance was well suited to the horizontal and vertical emphasis of modern architecture with its Cubist origins. Fashionable rooms were light and airy, lit by large picture windows. Wall and ceiling decoration was plain in style and curtains were straight drops with no valances or ties. The only pattern in a room was usually provided by the carpet or rug, scatter cushions and perhaps the drapery fabric. Geometric shapes were the most common. Open shelves and cupboards might display ceramics and plants, but otherwise paintings and ornaments were scarce.

· THE LEGACY OF POST-WAR AMERICAN DESIGN ·

Upholstery was not a prominent feature of this aggressively modern furniture. Many chairs had no textile covers at all with only strips of leather, canvas or some other coarse material slung across the frame or used to cover the thin padding of foam or rubber.

The use of latex foam as a substitute for traditional stuffing was universal. It had been discovered in 1928 and was cheap, it could be cut

*H*ere are some examples (right) of Eames and Saarinen's winning furniture at the exhibition of Organic Design in Home Furnishings, first held at the Museum of Modern Art in New York in 1941. Because of a shortage of raw materials, the chair legs were at first made in timber rather than the intended metal.

*A*lthough in some cases the entire molded body of a chair was fabric-covered, the chosen materials were usually rather dull; a plain color was necessary to emphasize the molded form of the over-all piece. On Saarinen's Tulip chair (left), the seat cushion is almost incidental.

or molded to shape and its fitting required little labor. Unfortunately it wasn't long-lasting and tended to lose its shape and even disintegrate with age.

The disposable, molded chairs were never very popular as antiques and so are not much in evidence today. However, they suited those who lived in small apartment blocks and developments where space-saving furniture was essential and stacking very useful. The idea of having living, dining and cooking areas combined in the one space was just coming into vogue and soon the demand was for more built-in furniture.

INNOVATION IN ITALY

significant contribution was made to post-war furniture design by Italy, a country which, although providing the source of inspiration for so much furniture over the centuries, itself came up with few innovative designs after about 1600. This sudden rebirth after the Second World War has been partly attributed to the willingness of Italian manufacturers to risk investment in new ideas, but has also owed much to the innate flair and conscious concern for style amongst young designers.

Italian designers in the immediate post-war period were more concerned with appearance than function and much of their furniture had, and still has today, a sculptural quality which

American products, with the emphasis on new materials and technology, have lacked. The same stick-like metal or pointed wooden legs were used for both chairs and sofas, but the upholstered bodies of the various pieces were essentially curved in shape and quite thickly padded with foam. Plain-colored and slightly textured or natural-looking fabrics were used as coverings.

During the 1960s, while designers in other countries were still concentrating on modular storage units, in response to demand for flexible furnishings, Italian designers were also devising modular seating systems in which individual seating units could be assembled in a variety of ways. Mostly consisting of fabric-covered blocks of polyurethane foam, these

units were generally available either single and armless, or with an arm on one side to form an end if required. This type of systems upholstery was popular everywhere in the 1970s and was found to be particularly useful for contract furnishing. One of the most admired ranges was Joe Colombo's *Additional System*, first produced in 1969. Colombo, who died young in 1971, was one of the first designers of the injection-molded all-plastic chairs that enjoyed such popularity in the late sixties.

Like many of his contemporaries, Colombo liked to use leather as an upholstery fabric, and leather has since become the hallmark of the more expensive luxury ranges of Italian upholstered furniture marketed so successfully throughout Europe and America.

The prime concern of Italian designers with style and technology is evident in this smart reclining chair designed by Osvaldo Borsani in the mid-fifties. Technical expertise is an obvious feature too.

The obsession of upholstered furniture designers with leather has led to some interesting sidelines. While not to everyone's taste, these bizarre chaises longues by De Sede (above) should get top marks for ingenuity.

A revival of chromium-plated tubular steel frames saw the production of numerous ranges of sofas and chairs with black or tan leather cushions (right). Outside Italy the combination was still not well received at a domestic level.

One of the most successful post-war Italian companies has been Cassina, for whom this marvelously inventive New York skyline sofa (left) was designed by Gaetano Pesce.

Italian designers moved quickly away from this sort of commercial fifties chair (below). The finish of the wood does not help the over-all design.

UTILITY FURNITURE

In the immediate post-war years Britain was at a tremendous disadvantage where industries of all kinds were concerned. Large expanses of the traditional centers of industrial production had been razed to the ground during the German bombing offensives and those factories which had survived were mostly equipped with out-of-date machinery and had in any case been geared towards the provision of armaments and airplanes. They were ill-suited to the manufacture of experimental furniture.

Although the general mood of the country hardly allowed for any yearnings for luxury, there was still considerable demand for new furnishings. Faced with the industrial situation and a shortage of raw materials, the Government maintained the restrictions on the production of furniture that it had enforced in 1942 and continued to allow only the manufacture of the limited standard ranges of cheap and simple, but well-designed, furniture available under the brand-name Utility. The employment of Gordon Russell as a design consultant to the Advisory Committee on Utility Furniture ensured that the principle of high standards of design available to the mass of the people was preserved.

Although Utility furniture was well made and well proportioned, it was too simple for most tastes and as soon as the production of other furniture was allowed after 1948, despite the imposition of high purchase tax, period styles soon re-appeared. What Utility furniture did achieve was the introduction into British homes of furniture in the modern style.

· THE NEW STYLES ·

Once the purchase tax on non-utility goods was lifted, more varied furniture and textiles became widely available. On the one hand antique-style furniture continued to be popular, especially for suburban middle-class homes. Textile manufacturers maintained their output of traditional floral cottons for the making of loose covers for three-piece suites and bedroom chairs, although they also produced modern fabrics for those with broader tastes.

When new furniture was in short supply, loose covers provided an ideal method of updating older pieces. An article in *House and Garden* in 1949 recommended a variety of materials for this purpose — a rayon plaid hopsack for a square-cushioned antique X-framed stool, a lime green and white small-patterned linen for a French rococo armchair, a plain textured linen for a wing chair and a rayon figured stripe for "traditional" pieces.

On the other hand young designers were beginning to experiment with new materials along American lines. The most successful of these was Ernest Race, whose cast aluminum side chair of 1947, with its upholstered seat and yoke, won him a gold medal at the *X Triennale* in Milan and subsequently found its way into more houses than any other individual piece of furniture of the period. Race also experimented with modern synthetic materials for upholstery. He not only used foam and rubberized hair for the stuffing, but covered the seats and backs of his chairs with fabrics such as Tygan, a cloth of plastic yarn which was advertised as being both spongeable and mothproof. It was Race's belief that it was the quality, not the period style of a piece that mattered most and that therefore modern designs could be happily combined with antiques in any interior. Not everyone agreed with him and most people preferred to see modern pieces in an appropriately

Gordon Russell's appointment to the advisory board on Utility furniture enabled him to further his strong belief that it was perfectly possible to produce low-cost furniture (left) without any loss of quality.

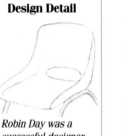

Design Detail

Robin Day was a successful designer. His polypropylene stacking chairs are still in use today.

British consumers preferred polished wood frames to the metal generally favored in America. This leather dining chair (right) designed by Robert Heritage is of a type still made for the contract market.

modern setting. Other successful designers of the period were Robin Day and Clive Latimer, who also won an international first prize at the Museum of Modern Art's 1948 Competition for Low Cost Furniture.

In 1951 the Festival of Britain was staged to give a boost to British design in all fields of the arts after the doldrums of the war years. The exhibits were mostly in a good honest modern style; there was not much that was remarkable for its originality.

· POPULAR UPHOLSTERY MATERIALS ·

Where upholstery was concerned, new three-piece suites were often still made in pre-war shapes and popularly covered in thick rayon moquette. Of rather large over-all size with deep seats and separate seat cushions, horizontal canted backs and broad arms higher at the front than the back, these cumbersome objects must have taken up a great deal of space in small suburban houses and the growing number of apartments which were springing up in and around the cities. More modern boxier and lighter armchairs and sofas were fashionably covered with broad striped or plaid materials, mostly of a rayon and cotton mixture.

During the early 1950s furniture in modern styles became much more widely available. The metal frames and laminated seats seen so much in America and Scandinavia were less popular with the British, who preferred light-colored beech frames with stick legs and box-shaped, foam-filled upholstery. Dining chairs were fashionably covered with washable plastic or plastic-based materials in plain colors, often with granular surfaces, and checks, plaids and stripes were still popular in living rooms. Tweeds and other coarsely textured natural-look materials became particularly common, the best and largest number being

Ernest Race was one of the most adventurous furniture manufacturers. His designs, such as the one above, made Race Furniture a leading name in the post-war period. Over a quarter of a million of his BA cast aluminum chairs were sold during the 23 years it was in production.

Another British designer who became a household name in the fifties was Robin Day. This design for a living room (right) for Hille has come to typify design for its date — 1957. The simple foam-filled cushions are covered in plain textured material and the insignificant straight metal legs are both elegant and functional. The deep pile carpet, plain walls and curtains, and lack of clutter were still rather a new trend and difficult to sustain in the home.

*S*ince the war Heal's has been one of the foremost promoters of contemporary textile design. These furnishing fabrics (above) from a 1952 catalogue are absolutely typical of the period in both color and pattern.

*T*he living area in this fashionable 1957 flat (right) has many fashionable features; a pine-panelled ceiling, white walls, a large picture window, a wood-framed settee with loose foam cushions, molded chairs.

produced by the Donald Brothers in Scotland under the name Old Glamis.

Colors generally were rather harsh and unsubtle and clashing combinations were not uncommon. Lemon yellow, lime green, purple, grey, black and white and deep red were the most popular.

· INTERIOR LAYOUTS ·

As in America rooms were sparsely furnished. Light-colored veneered cabinets tables and hi-fi units and open shelves displaying a small number of books, plants and ornaments, were placed against plain white or cream walls. Curtains and rugs were patterned with abstract designs in bright colors, often on a white ground.

By 1960 interior design had generally become less brash. Colors were softer — blues, greys and pinks were quite popular – and more harmonious combinations were used. The various lines of sofas and chairs no longer radiated outwards, but were right-angled and legs were straight. Upholstery fabrics were still predominantly textured and plain in color. Velvet began to make an appearance, as did leather or "mock leather" (PVC). Most upholstery was completely flat and even, without any form of trimming or piping, although some pieces had shallow-buttoned areas.

The 1960s also saw the introduction of natural materials to interiors. Polished timber flooring and rough, exposed stone and brick walls and hearths were complemented by textured fabrics in natural earthy colors, such as brown, beige and terracotta, and by tan and black leather. Scandinavian oiled teak settees and armchairs were particularly fashionable and were a strong influence.

In complete contrast to the natural look, which was chiefly advocated by architects, some manufacturers adopted the more folk-oriented aspects of Scandinavian design and used lighter, brighter colors for upholstery, such as basic red, blue and yellow set against white walls and plain colored or striped curtains. Towards the end of the decade orange became a great favorite.

A great many designers tried their hand at producing molded chairs. In this design by Conran

(below) extra strength has been provided by forming the legs from one continuous length of metal.

124

QUALITY OF DESIGN

The high standards of design which Scandinavian furniture producers of the 1920s and 1930s had established was maintained in the three decades following the war. Many Scandinavian designers were also architects who made furniture to suit their living spaces. The majority of chairs and sofas were still made with wooden frames, although foam-filled upholstery generally replaced the use of laminates for seats. As elsewhere black leather was the most popular, if the most costly, material, but wool and other plain woven fabrics were also used. After 1950 natural-look textured fabrics were favored and brighter colors and plaids were in vogue.

Just as Scandinavian design influenced American manufacturers, so American trends were adopted in Scandinavia and a number of designers began to work with metal. Most notable of these was the Danish designer Arne Jacobsen, whose smoothly upholstered "Swan" and "Egg" chairs have been very successful. Influential designers working in wood were Finn Juhl and Hans Wegner.

The oiled teak settee with tan leather upholstery (below) by the Danish designer Finn Juhl provided the inspiration for much sixties furniture throughout the world.

Arne Jacobsen's Egg chair (right) is perfectly balanced. Stretching the leather over such a large curved expanse demands great skill.

Design Detail

A hammock chair designed by Poul Kjaerholm, made from cane, has a leather cushion as the only upholstery.

THE PRESENT DAY

PROGRESSIVE DESIGNS
·
CONTEMPORARY INFLUENCES
·
RETURN TO TRADITION
·
MODERN TEXTILES
·
THE RANGE OF CHOICE

*A totally new concept of the
twentieth century*

PROGRESSIVE DESIGNS

urniture design in the last 25 years has developed at two distinct levels. The most advanced designs have largely been successful in the contract market and an identifiable international style has evolved. The domestic market, on the other hand, has undergone a series of rapid changes resulting in a greater number of eclectic interiors than at any other period in history. Co-ordination of color and pattern has become more important than consistency of style.

Despite some wonderfully inventive and striking furniture executed principally in synthetic materials, truly modern designs have seldom been enthusiastically received by the ordinary consumer, possibly because the

*T*his table and chairs (below) designed in 1979 by Stefan Wewerka could perhaps be viewed more as a work of art than as furniture. They are covered in a Bauhaus fabric.

*E*ero Saarinen's space-age Globe chair (right) appeared in a number of films in the late sixties, but it needs courage to accommodate it in a domestic room setting.

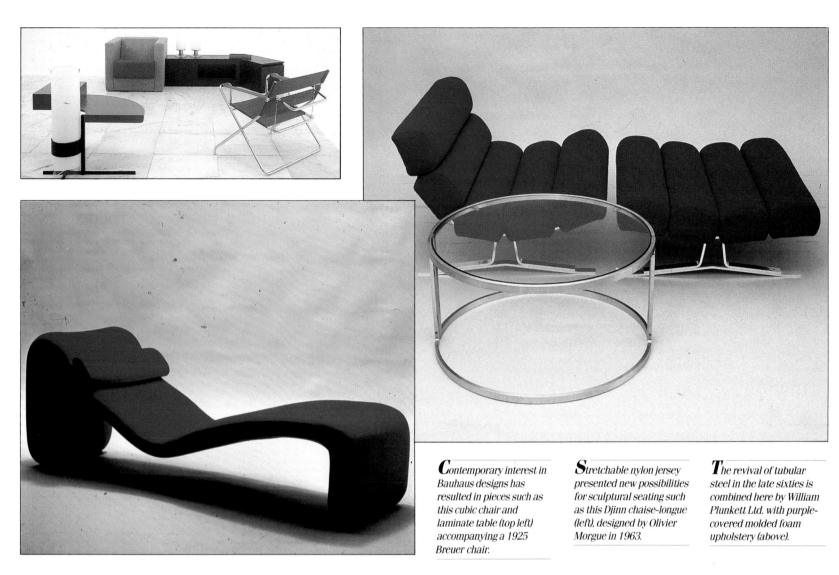

*C*ontemporary interest in Bauhaus designs has resulted in pieces such as this cubic chair and laminate table (top left) accompanying a 1925 Breuer chair.

*S*tretchable nylon jersey presented new possibilities for sculptural seating such as this Djinn chaise-longue (left), designed by Olivier Morgue in 1963.

*T*he revival of tubular steel in the late sixties is combined here by William Plunkett Ltd. with purple-covered molded foam upholstery (above).

smooth surfaces and precise outlines require a complementary sparse background and make no allowances for ordinary household clutter. The austere appearance of starkly modern furniture fails to give the impression of warmth and comfort that the majority of people want in their homes. By contrast, in a commercial environment, particularly in this high-tech age of computerized office equipment austerity of design gives an air of businesslike sophistication and efficiency.

The most progressive designs of the 1960s relied heavily on a combination of metal and leather, although plastics and fiberglass replaced laminated wood for seating, and glass was widely used for flat surfaces on other types of furniture. Tubular steel tended to be thicker than before. Surprisingly, modern tubular steel

furniture enjoyed only a short-lived popularity, while reproductions of many of the original 1920s Bauhaus designs, which had provided the source of inspiration for them, became far more fashionable and are still bought for high-tech office and domestic interiors today.

This form of revived modernism was soon challenged by a number of futuristic designs executed in plastic and fiberglass with a smooth, glossy, white finish. A stimulus was provided for this type of furniture by the space race and a consequent interest in science fiction. Viewed retrospectively, the most famous of these designs was Aero Aarnio's "Globe" chair manufactured by Asko Finnternational. This was basically a white, fiberglass sphere, supported on a white-enameled circular aluminum base, with a slice cut off one side to reveal a

smoothly upholstered interior with a thick, loose cushion.

Upholstery on all kinds of furniture was almost exclusively of molded foam and was still most fashionably covered in leather, with nylon jersey, a fabric that could be stretched evenly to give a smooth fit without obvious joins over virtually any shape, providing a cheaper alternative.

Throughout the 1960s and into the 1970s, plain white walls were a popular background for all modern furnishings. Modern prints and paintings provided some color and decoration, but otherwise wall surfaces were completely bare. In old houses, fireplaces, picture rails and even cornices were removed, and spotlights and spherical Japanese paper lampshades replaced conventional light fittings.

CONTEMPORARY INFLUENCES

The most pervasive influence on modern furnishing in Britain has been Terence Conran's Habitat store, opened in the Tottenham Court Road in London in 1964. Initially few designs were totally original — Bauhaus reproductions, Thonet bentwood designs and various directors' and deck chairs were all popular — Habitat could supply simple, modern furnishings of all types from crockery to bed linen and a complete "Habitat" look could be achieved at astonishingly little cost. Still highly successful, they have maintained their impression of simplicity and clean lines. Early Habitat sofas and armchairs had light-colored wood frames of Scandinavian appearance and deeply stuffed and sometimes shallow-buttoned plain, loose cushions. In the early seventies these were followed by a range of painted, tubular, metal-framed chairs and sofas with matching cotton cushions, in the primary colors that were particularly fashionable.

For those who found primary colors too garish, there was a choice of natural colors such as beige, oatmeal and brown, and textured fabrics, canvas and corduroy were also used a great deal. Sofas often came in units. Traditional sofas were quite boxy in shape with square-ended arms and straight, flat seats.

In the early 1970s two-seater sofa-beds were introduced. These have since proved to be popular and particularly convenient for small houses and apartments.

· ECLECTICISM ·

During the late 1960s and early 1970s ethnic furnishings were fashionable, particularly amongst young people. Imported ethnic rugs, wall hangings and other furnishings had the advantage of being extremely cheap. An otherwise bare room could be transformed with brightly colored patterns. The fashion brought with it a trend for floor-level seating and enormous cushions, and mattresses covered with Indian bedspreads replaced conventional seating. In 1969 an Italian design team in Milan came up with the "Sacco," a large bag of plastic granules that served the same purpose and was naturally molded to a comfortable shape by the sitter.

Another London store that had a consider-

The idea for the cover of this Sindbad chair and ottoman (above), by the Italian designer Magistretti, was inspired by a horse blanket he saw in a London shop. Other designers too have used rugs as "throwover" upholstery.

The tradition for laminated furniture has been maintained by contemporary designers. As this bent birch chair by Esko Pajamies (right) shows the upholstery is more sophisticated.

The Sacco (above) required some agility from the user, not just because it was difficult to vacate. The polystyrene granules move with any movement.

Earthy colors provided a suitable background for ethnic textiles. Beds were covered with rugs and bedspreads and piled high with cushions (below), not unlike the Moorish divans of the last century.

In the early seventies the use of primary colors was promoted by a number of companies. This cheap and cheerful look (right) was well suited to both the pockets and taste of a new generation.

The Habitat look is now instantly recognizable. White paint and bare polished wood have remained its chief characteristics. This page from a 1977 catalogue (below) shows just one of their many systems ranges of foam seating units.

able influence on furnishing styles — albeit unintentionally — was the boutique *Biba*, situated in London's Kensington. The brown and plum-colored shop fittings, with long-fringed Edwardian satin lampshades, bentwood coat stands, Victorian *jardinières* and art nouveau wallpapers, stimulated an interest in the late nineteenth and early twentieth centuries. Collecting Victoriana became the rage, and spoon-back chairs, *chaises longues*, chesterfields and simple balloon-back dining chairs rocketed in price. Dralon velvet was suddenly in great demand and quickly became popular for seating in all styles.

To complement the furniture, William Morris designs were reproduced in a wide variety of original and modern colorways on both wallpapers and furnishing fabrics. Victorian bamboo, with panels of lacquer or woven grass, was also widely collected and, before long, imported cane furniture was being sold everywhere for use inside and out.

131

POP ART

Technological advances in industrial materials gave rise to a number of less serious furniture designs that have since been included in the category of Pop Art. Best known are Peter Murdoch's folding paper chair (British-designed, but American-made), and the first inflatable Blow chair (below), produced in Italy by Zanotta of Milan, who also produced the Sacco chair. Although not exactly upholstery, inflatable chairs were at least based on stuffed seats and were intended to provide the same degree of comfort and support.

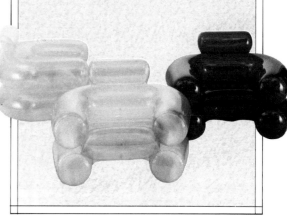

RETURN TO TRADITION

As the supply of English Victoriana ran out — most of it having gone abroad, a great deal to America — those who could no longer afford antiques turned to stripped pine furniture as a cheap alternative. The "country look" that this gave to rooms was offset by ranges of small-patterned fabrics and wallpapers, many with sprigs of flowers in white or cream on a darkish ground such as those produced by Laura Ashley, now an international company. As pine-framed seating was not traditionally made, ordinary upholstered sofas were considered suitable accompaniments in these pine-furnished interiors, and rounded shapes reappeared.

These trends really marked the beginning of the return to traditionalism which had been the prominent feature of fashionable interiors for the previous ten years. Throughout the entire twentieth century there has always been an underlying trend in all countries for the "antique" look, and particularly for eighteenth- and early nineteenth-century styles. In France, tapestry, needlework, damask and velvet have been consistently popular for both antique and reproduction furniture. In Britain, floral chintz loose covers are still common. Rather bland, conventional colors, such as pale blue, green and gold, are most popular. In conventional rooms, carpets, curtains and walls are generally plain and the color and decoration are provided by ornaments in the form of clocks, picture frames and pretty china.

Although today's fashionable interiors are also traditional, there have been three notable differences in recent years. First, a much greater interest has been shown in authenticity; second, the use of color has been much bolder; and third, textiles have reasserted themselves as the most dominant feature. Curtains particularly have become more elaborate, festoon curtains being especially popular, although usually inappropriately used, and elaborate drop curtains with fancy draped valances and matching tie-backs and tassels have reappeared. Pleated or gathered skirts on both fixed and loose chair and sofa covers and frilled cushions are made in matching materials. In bedrooms, frills and flounces are much in evidence and lace and satin are used for bedcovers and cushions.

Three-piece suites, as such, have gone out of favor and although sofas are still universally used, chairs are frequently of varying types and styles and may even be covered in a fabric with a different, yet complementary, pattern.

The French rococo armchair (below) has become an internationally popular classic. It looks particularly good with a small-patterned fabric.

This eye-catching pattern with its representations of oriental procelain gives prominence to an otherwise simple chair (above).

The wonderfully strong colors and bold stripes blend perfectly with the natural wood finishes in this "antique" interior (right).

MAKING SCATTER CUSHIONS

Scatter cushions can alter an interior quickly and cheaply and brighten up a plain sofa or chair. The edges can be plain or frilled, finishing with cord or piping.

Make the piping first (see page 110) and cut out the cushion pieces using the cushion pad to determine the size. Allow 12mm (½in) all around for the seam allowance. Pin the piping all around one of the pieces on the right side, with the raw edges facing outwards. Clip the

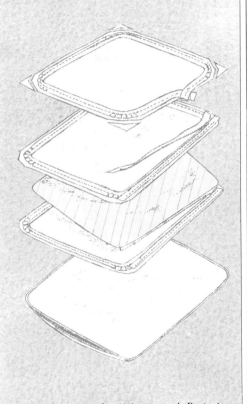

corners to ease the piping around. Baste in place. Open the zipper and place one edge against the piping. Stitch with the zipper foot as close as possible to the piping. Position the other side of the zipper in the corresponding position on the other cushion piece. Do not sew across the ends of the zipper. Sew the two cushion pieces together and turn right side out through the zipper opening.

Simple mid-eighteenth-century-style chairs (right) have been enlivened by some imaginative trimmings. The decorative brass nailing is a particularly skilful device and has obviously been inspired by the designs of George Hepplewhite.

In a traditional room in a large London flat (below), the formal Georgian furnishings have been cleverly combined with soft pastel shades to produce an interior of considerable warmth.

133

lazed chintzes with mostly traditional floral and chinoiserie patterns have been popular in brilliant colors such as turquoise, emerald green and deep pink. Attention is drawn to edges by the use of contrasting colored piping and trimmings. Classical colors such as black, terracotta and gold have recently been reintroduced. For those who prefer a softer look, more conventional fabrics, many with small sprigs of flowers, or diamonds or other small geometric patterns, are also manufactured in pastel shades.

Because of the interest in authenticity, textile manufacturers have produced greater numbers of fabrics either directly copied from, or inspired by, old documents. France has maintained the centuries-old tradition of fine-quality textiles where both design and manufacture are concerned. In addition, reproduction furniture has seen a considerable

The combination of a simple broad check with a more formal traditional pattern emphasizes the country look (right).

The unusual shape of these pieces (below) has been cleverly outlined by the fabric choice — a plain, dark color and a striking horizontal stripe.

The sophisticated sheen and subtle colors of these modern American fabrics (bottom) are typical of the taste for glamour.

134

increase in demand, mostly because the price of antiques has risen so sharply, and better and far more varied styles are widely available. It is now possible to buy either authentically reproduced pieces ready-covered in appropriate fabric, pieces with calico covers ready for the top fabric of your choice, or even chair frames for do-it-yourself upholstery, a practice which is becoming increasing popular.

Naturally not all interiors are centered around antiques and there are signs of a reaction against them. Glossy-finished fabrics with abstract patterns, in muted tones of gray, mauve and pink, have become quite numerous and "luxury" appears to be a new theme.

For a tiny minority there has been a craft revival, although makers have been more interested in the nature of the wood itself and in organic shapes than in the upholstery or function. Young designers of upholstered furniture within the more commercial market, perhaps inevitably, are still struggling with traditional forms of seating, but attempting to create a new look by angling them in different ways.

The availability of foam has undoubtedly led to a far greater range of shapes than is possible with conventional methods of upholstery. It is also much cheaper. However, recent legislation against its use because of fire hazard may have a restricting influence in the future.

Scatter cushions (left) have been much in favor during the past 10 to 15 years, particularly in bedrooms. Fancy borders and decorative trimmings have become essential to their design.

Brilliant colors have enlivened a traditional cotton fabric on this simple Victorian armchair (below). Its combination with simple striped and checked fabrics in matching colors gives a natural freshness to this room.

A SHAPED CUSHION

Cushions can be shaped or made from quilted, patchwork or appliqued fabrics. To make a circular cushion, you will need a cushion pad to determine the diameter of the finished cushion. Cut out a paper pattern by folding the paper into quarters and with a pencil and string the length of the diameter, draw a circle. Cut out through all widths. Now cut one piece of fabric using this

pattern, remembering to center any prominent motifs and allowing 12mm (½in) seam allowance. Cut the paper pattern across where the zipper is to go. Move the pattern pieces apart by 3cm (1in) and cut out the two pieces with the seam allowance for the zipper. Insert the zipper (see page 65). Make piping (see page 110) and attach it (see page 133). Join the pieces, turn right side out and insert the pad.

The greatest advantage of the present situation is its diversity. Not only is it possible to buy upholstered furniture of virtually any date or style, but is quite acceptable to have a mixture of styles.

The trend during the eighties for a post-modernist approach to architecture, in which classical ornament is imposed on basically modern structures, has been extended to interior design, particularly in America. Morris Lapidus' glamorous and sophisticated designs for Florida hotels have been especially influential and combine modern streamlined fittings with decorative antiques in a manner reminiscent of the opulent hotel and liner interiors of the thirties, but with a stronger grasp of color and form. On a domestic level, the same decorative pieces may also be combined with modern art works.

Such self-conscious post-modernism is really only a reflection of the natural evolution of the contemporary home. Against a background of technically efficient and sometimes well-concealed modern conveniences, most households simply collect an assortment of furnishings from a variety of sources, some passed down from previous generations of their family, some bought new and some simply second-hand, as yet without any antique value.

· THE CORRECT SHAPE ·

Unlike other types of furniture, the appearance of upholstery can be easily changed, and there is now a more exciting range of fabrics and trimmings available than at any other time. Although a piece of upholstery will benefit greatly from correct shaping, the actual cover doesn't have to be exactly copied, and striking effects can be achieved by using totally different patterns.

A well-known London decorator writing in the thirties gave some sound advice which is equally relevant today. "There is no doubt that modern fabrics are adaptable; quite frequently one sees old chairs of classical type covered with very contemporary fabrics, and the effect is charming, but I have yet to see a strictly modern chair upholstered in tapestry or brocade of classical design that would prove anything but dreadful." Certainly something to bear in mind.

White piping has been used in stark contrast to the black upholstery to emphasize the striking lines of this thirties revival chair (left).

While upholstery has not always been a prominent feature of twentieth-century furniture, the right choice of fabric, however simple the shape it covers, can make all the difference (above).

The futuristic design of this contemporary American chair (right) proves that there is still plenty of potential for innovative design.

A large contemporary chair (right) can best be viewed as a piece of sculpture. By the designer George J. Sowden, the chair displays a printed cotton fabric draped over the arms.

*N*ot dissimilar items in terms of basic shape, the more conventional sofa (right) has been upholstered in a starkly modern fabric with the added interest of a visual joke. It needs plain surroundings to appreciate the two figures already sitting on the seat.

*T*he winter room (above) is one of the interiors in Thematic House in London, devised by the post-modernist Charles Jencks. He designed the armchairs; everything in the room has a symbolic meaning.

*O*riginality is sometimes difficult to achieve without losing sight of function. The marvellous pleated and draped upholstery of this sofa (right) looks marvelous now; one wonders what it looks like when it has just been used.

THE PRESENT DAY

GLOSSARY

ARTS & CRAFTS — an artistic movement dating from the last 40 years of the nineteenth century which advocated a return to hand craftsmanship and simple design and construction. Its most famous member was the founder William Morris. Its influence was strongly felt in America.

ART NOUVEAU — an asymmetrical, curvilinear style with elongated naturalistic forms as its chief characteristic. Popular in France and Belgium around 1900, the name derives from a fashionable gallery, the *Salon d'Art Nouveau*, in Paris.

BACK-STOOL — usually referred to now as a farthingale, it was the most popular type of chair in seventeenth-century Britain and Europe. It was basically a square stool with stretchers and a low upholstered back.

BAROQUE — a late seventeenth-century style originating in Italy. Its chief characteristics were exuberant decoration and symmetrical curved lines, in particular the G-scroll.

BENDED-BACK CHAIR — curved for comfort by means of a baluster splat situated at shoulder level.

BIEDERMEIER — a simple, functional, essentially middle-class style prevailing in Germany and Austria during the first half of the nineteenth century and based on French Empire.

BOLSTER — a long, round-sectioned cushion; originally a pillow, but later adapted for use on sofas, daybeds and chaises longues.

BRAID — a woven band of threads used for decorative purposes on other textiles.

BROCADE — a woven material with raised decoration of gold or silver thread.

BROCATELLE — an apparently silk fabric looking similar to damask with a silk weft but a linen warp. Cotton and wool were sometimes used instead of silk. Large-patterned in the seventeenth century, but more varied in the nineteenth.

CABRIOLE LEG — a leg of reversed S-shape which originated in China and was popular in the West from the beginning of the eighteenth century onwards.

CAFFOY — a term which appears in early inventories and refers to a woolen fabric of some kind.

CALICO — a plain cotton cloth originating in India.

CAMLET — a ribbed woolen cloth which was sometimes treated in some way after weaving to produce a pattern.

CASE OR SLIP COVERS — fitted loose covers, usually made in a simple, inexpensive material, to protect more expensive fixed upholstery when the furniture was not in use.

CHENEY — a woolen cloth like camlet; sometimes patterned.

CHINOISERIE — a European interpretation of Chinese and Japanese decorative art and architecture. First popular in the seventeenth century.

CHINTZ — a glazed cotton cloth with various designs printed in colors.

CLAW-AND-BALL FOOT — a popular form of carved foot found on eighteenth-century chairs. Thought to originate in China; literally in the shape of a claw clutching a ball.

CORDUROY — a woven fabric with a pronounced warp-wise rib.

CRETONNE — a printed cotton cloth slightly heavier than chintz and without the glaze. Popular in the late seventeenth and early nineteenth centuries.

CREWELWORK — a particular type of embroidery executed in fine-spun wool on a plain linen or other ground, usually imitating seventeenth-century imported Indian printed cottons.

138

CURULE — a classical Greek stool of X-frame construction.

DAMASK — a mostly single-colored material with the pattern woven in a contrasting weave so that light is reflected off it in different ways.

DIMITY — a strong cotton fabric woven with a variety of repetitive patterns such as stripes, checks, diamonds, herringbone and sometimes flowers. In England used mostly for clothing, but in America for furnishings too.

EBONIZED WOOD — sometimes wood was veneered with real ebony (a hard black wood), but usually the term refers to a light-colored wood stained black and polished to imitate ebony.

FARTHINGALE CHAIR — called a farthingale because it could easily accommodate the hooped farthingale skirts fashionable in the seventeenth century (see back-stool).

FOLKWEAVE — a loosely woven fabric made from coarse yarns. Popular from the late 1930s onwards when it mostly had plaid or striped designs.

FUSTIAN — a thick, strong, coarse cotton and linen cloth, often used as lining for bed curtains.

GALLOON — a decorative braid woven from gold, silver or silk threads.

GENOA VELVET — a Genoese woven silk fabric with a large pattern in silk pile against a plain background. Very popular in seventeenth- and eighteenth-century Europe, particularly in red and green on a white ground.

GIMP — a narrow woven braid used as a border to cover tacks and raw edges of material.

HAIRCLOTH — a fabric woven with washed and combed horsehair combined with linen or other threads. Introduced in about 1750 it is now available in a much wider range of colors and patterns, but is rarely used (see horsehair).

HARATEEN — a woolen fabric on which a pattern has been "hot-pressed" — printed with hot plates pressed against a carved wood-block.

HESSIAN — a plain woven fabric made from jute and used for traditional upholstery to support or cover stuffings.

HORSEHAIR — curled horsehair was used for the stuffing of chairs from the seventeenth century onwards (see haircloth).

JAZZ MODERNE — a term used in the 1920s and 1930s to describe the modern designs of cubist inspiration which characterized the French furnishings shown at the 1925 Paris *Exposition des Arts Decoratifs et Industriels*.

JUGENDSTIL — an artistic movement of the 1890s based in Munich and promoting modern design.

KLISMOS — a classical Greek chair much copied throughout Europe around 1800. It has saber legs at both the front and back and a broad yoke (top rail).

KNOTTING — a form of decoration on fabric achieved by stitching lengths of thread, which had been previously knotted on a special shuttle, to a prescribed pattern.

LACQUER — a form of surface decoration originating in China and Japan and achieved by the application and polishing of numerous coats of the gum of the lac tree. European imitation of lacquerwork was called japanning.

LUSTRING — a light silk fabric which was treated with gum to give it a high gloss (also lutestring).

MARLBORO LEG — a form of straight chair leg popular after the middle of the eighteenth century. It is usually chamfered (flattened off) slightly on the inner corner.

MOQUETTE — a similar fabric to velvet with a wool pile on a cotton ground. Modern moquettes usually have a silk and rayon mix pile.

NEO-CLASSICAL — a style of architecture and decoration popular in Europe and England from the 1760s onwards and slightly later in America. Classical Greek and Roman motifs were applied to contemporary forms of furniture.

ORMOLU — a metal used extensively for furniture mounts from the seventeenth century onwards in which gold leaf was fused to a bronze or brass base (sometimes known as gilt-bronze; in France *bronze doré*).

PALLADIANISM — an early eighteenth-century style based on the earlier style of the Italian architect Andrea Palladio.

PARAGON — a stronger form of camlet.

PASTICHE — an attempt to design something "in-the-style-of" an earlier period without trying to reproduce it exactly.

PLUSH — a fabric similar to velvet, but with a long pile. It has been made of silk, wool or cotton. It was particularly popular in the nineteenth century.

PLYWOOD — a board produced by glueing sheets of veneer together with the grain of alternate sheets running in opposite directions. Used extensively in the twentieth century because it is both strong and cheap to produce.

PRIE-DIEU (or devotional chair) — originally a kneeling chair used for religious purposes, but later a low-seated chair with a tall back and a padded back rail. Very popular during the second half of the nineteenth century in France and Britain.

139

PROGRESSIVE — a popular term now used to describe "modern" designers of the late nineteenth and early twentieth centuries.

RAYON — a synthetic fabric used extensively for a variety of purposes from the 1950s onwards.

RENAISSANCE MOTIFS — the decorative motifs based on the classical orders of architecture and antique Roman ornament which were revived in Italy (and subsequently elsewhere in Europe) in the fourteenth century.

REPP — a plain fabric with a pronounced rib running from side to side.

ROCOCO — a decorative style of the first half of the eighteenth century derived from the baroque, but lighter, and characterized by asymmetrical arrangements of scrolls, foliage, shells, icicles and rockwork.

SABER LEG — a slightly concave leg used in classical Greece and revived in Europe around 1800. Very popular for English Regency and Biedermeier furniture (also called a scimitar leg).

SATIN — usually refers to a shiny-surfaced silk, actually any fabric with a very smooth surface.

SECESSIONISM — an Austrian and German movement of the early twentieth century based on the principles of the British Arts and Crafts movement, but aimed at industrial production.

SERGE — a woolen material woven to produce a diagonal ribbed pattern.

SHALLOON — a type of serge.

SLIP COVERS (see case covers).

SPLAT — the central upright section of the back of a chair; can be flat or have openwork carved decoration.

SQUAB — a loose padded seat cushion invariably used on caned seats. Originally seen on "squab frames," low carved stools which, in the seventeenth century, were an essential part of any grand suite of furniture.

SYSTEMS FURNITURE — ranges of furniture consisting of a number of separate units which can be fitted together in a variety of different ways.

TABBY — a ribbed silk that sometimes had a pattern.

TABOURET — the French term for a stool. Stools, rather than chairs, were a popular form of seating for French women in the seventeenth century and were an indication of their status.

TAFFETA — a silk with a plain weave.

TAPESTRY — a cloth into which colored threads were hand-woven to produce a pattern. Now woven by machine. Originally used for wall hangings, but from the seventeenth century onwards for upholstery and cushions too.

TINSEL — a silk material with gold threads woven into it to make it sparkle. Many examples of this type of material survive from the seventeenth century.

TOILES DE JUOY — the popular term for French printed cotton chintzes. Printers working around the town of Juoy in the eighteenth century were famous for the high quality of their work.

TRANSITIONAL — term used to describe furniture displaying the characteristics of two styles at once. Usually it refers to French pieces dating from the 1750s and 1760s which have both rococo and neo-classical features.

TURKEYWORK — the European woven imitation of Turkish rugs. At first designs were Near Eastern in style, but were soon replaced by floral patterns.

TWEED — a heavy fabric woven mainly or wholly from wool.

UTILITY — British furniture made under the *Utility Furniture Scheme* between 1942 and 1952. Restricted all production to a government-specified range of designs and prices, following a shortage of raw materials.

UTRECHT VELVET — the wrong term used to describe high-quality woolen velvets made in Holland from at least the seventeenth century onwards. Velvets were not actually made in Utrecht itself.

VELVET — a rich pile fabric used extensively for upholstery at all dates. Usually silk, and less commonly wool, but could also be cotton. Today rayon is used.

WEBBING — a band of coarse woven fabric stretched across a seat frame to support the padding.

WORSTED — a collective term used to describe any woven fabric made with woolen yarn or thread. It usually implies the use of a long, combed wool fiber.

140

INDEX

141

142

143

CREDITS

Quarto would like to thank the following for their help with this publication and for permissions to reproduce copyright material.

Key: A — Above; B — Below; R — Right; L — Left; F — Far;
C — Center; T — Top; Bt — Bottom

Norman Adams Ltd: pp42(R), 43(L)
The American Museum in Britain: pp35(B), 68
Apter Fredericks Ltd: pp54(A), 55(A), 90(BL)
Aram Designs Ltd: pp104, 105(AR,BR), 107(A,C)
Barling of Mount Street: p24(L)
Arthur Brett & Sons Ltd: p15(AL)
The Bridgeman Art Library: pp11, 12(L), 26, 28(A), 29(L), 30(B),
    32, 34, 35(A), 38(R), 44(B), 45(A), 46, 48(B), 25(B), 53, 56(L,R),
    57(L,R), 58, 60(AL), 61(L), 62(L), 36(L), 64(AR,BR), 69, 70, 72,
    73(AL,B), 76(A), 78(A), 79(AL), 81(AL,AC,AR), 82(AL,BR),
    85(BL), 86, 87(AR), 88, 96, 97, 98(L), 100, 101(BL), 109(AL),
    113(AR)
Angela Burgin: pp73(AR), 76(B), 78(BR), 87(AL), 99(A)
Rupert Cavendish: pp60(AR), 66(A,B), 67(L)
Chanteau: pp18(L), 22(B), 50(B), 51(AR)
Christie's Colour Library: pp3, 19(A), 27(B), 31(L), 38(L), 39(L,R),
    44(A), 51(L), 61(R), 91(R), 93(BR), 106(A,B), 108(L), 121(AL,BR)
Colefax & Fowler: pp133
The Design Council: pp74(B), 80(BL), 83(BL), 84(A,B), 85(AL,BR),
    93(AL), 94, 95(AL,AR,BL), 101(AL), 105(AL), 109(AR),
    122(AL,BR), 123(AL), 124(AR,BR), 129(BL,AR)
The Designer's Guild: pp19(BL), 135(B)
Derwent Upholstery: p5
E.W.A: pp15(AR), 18(R), 49, 75(A), 118(L), 126, 130(A),
    131(AR,AC), 133(B), 134(CL)

Gainsborough Silk Weaving Co. Ltd: pp16(BR), 40(AR), 51(BR)
Habitat: pp15(B), 75(B), 131(BL)
Heal's: p124(AL)
Angelo Hornak: pp36, 80(AR), 102
Kingcome Sofas: pp19, 132(L)
Knoll International: p136(BR)
Parker Knoll: p14(A)
William L Maclean: pp79(AR), 87(BR)
Monkwell: p77(CR)
Osborn & Little: pp132(R), 135(A)
Pallu & Lake: p132(C)
Percheron: pp17(AL,BL), 23(L), 25(R), 134(AR)
Sahco-Hesslein UK Ltd: p17(AR)
Sanderson: p74(A)
Sekers: p134(B)
Peta Smyth — Antique Textiles: pp16(AR), 23(AR,BR), 30(A),
    40(CR), 45(B), 48(A), 52(AL), 60(C,B), 77(CL), 90(T,B), 93(C)
Sotheby's: pp12(R), 42(L), 63(R), 64(BL)
Stuart Renaissance Textiles: pp12(C), 17(BR)
The Victoria & Albert Museum: pp8, 13(A), 14(B), 16(AL), 20,
    24(R), 27(A), 28(B), 29(R), 31(C,R), 33(L,R), 41(R), 50(A), 52(AR),
    55(B), 62(R), 64(AL), 77(F,R), 79(BR), 81(BR), 82(AR), 90(A,Bt),
    91(L), 93(AR,BL), 95(BR), 99(B), 115(L), 125(BL,AR), 136(AR)
Watts: pp77(CT), 78(BL)
Joanna Wiese: p113(CL,BC)

Every effort has been made to trace and acknowledge all
copyright holders. Quarto would like to apologize if any
omissions have been made.